Aristotle's Classification of Animals

PIERRE PELLEGRIN

ARISTOTLE'S CLASSIFICATION OF ANIMALS

BIOLOGY AND THE CONCEPTUAL UNITY OF THE ARISTOTELIAN CORPUS

TRANSLATED BY
ANTHONY PREUS

UNIVERSITY OF CALIFORNIA PRESS
BERKELEY · LOS ANGELES · LONDON

Originally published as *La Classification des animaux chez Aristote: Statut de la biologie et unité de l'aristotélisme,* © 1982 Société d'édition "Les Belles Lettres." Revised edition.

University of California Press
Berkeley and Los Angeles, California

University of California Press, Ltd.
London, England

Library of Congress Cataloging-in-Publication Data

Pellegrin, Pierre.
   Aristotle's classification of animals.

   Translation of: La classification des animaux chez
Aristote.
   Rev. ed. produced through collaboration between
P. Pellegrin and A. Preus.
   Bibliography: p.
   Includes indexes.
   1. Zoology — Classification. 2. Zoology —
Classification — History. 3. Aristotle. I. Preus,
Anthony. II. Title.
QL351.P4413 1986   591'.012   85-23305
ISBN 0-520-05502-0 (alk. paper)

Printed in the United States of America

1 2 3 4 5 6 7 8 9

A toutes les bêtes . . .

CONTENTS

TRANSLATOR'S NOTE

La Classification des animaux chez Aristote was sent to me, as required reading, prior to a conference on Aristotle's biology in 1983. It seemed to me that this book would revolutionize the interpretation of Aristotle's biological works, for its thesis, though radical, made many things clear that had been persistently obscured by all previous discussions of those works. I also thought that the book could have the more general effect of making necessary a general revision of the interpretation of the Aristotelian terms *genos* and *eidos,* even in logical and metaphysical contexts, for the book appeared to refute all attempts to translate those terms as "genus" and "species" as those words are understood today. I looked forward to meeting the author of such an important book.

Pierre Pellegrin turned out to be a young, witty, and amiable scholar. Others at the conference had much the same impressions of the book and its author as I had. Alan Gotthelf, the organizer of the conference, asked whether I might be interested in translating *La Classification* into English, since he was happy with my translation of an essay by Paul Moraux. When I expressed some reluctance at taking on such a large project, Ernst Mayr stepped in and persuaded me that the book ought to be translated, and that I was the one to do it.

The translation eventually became a revised edition, produced through collaboration between Pellegrin and me. At first we collaborated by mail, but when a complete draft was ready, we were able to spend several intense weeks going over the entire volume. We tried to make the translation accurate, and we tried to recreate something of Pellegrin's French style in the English version; we also corrected some errors, adjusted emphases, clarified the argument, and introduced some additional evidence for the major theses. We even eliminated some pages that no longer seemed to fit.

I thank Dennis Schmidt, Cynthia Freeland, and Alan Gotthelf for finding several remaining gallicisms and outright errors. Doubtless some infelicitous renderings remain; I take responsibility for them.

Anthony Preus

PREFACE

Originally this work was intended to be a short article on Aristotle's animal taxonomy. Sharing common prejudices, I expected to deal with the problem quickly, having noted some dark areas left by Aristotle in his construction of a "natural" classification of living things. But my research grew by negating the assumptions upon which it was originally based; I was seduced, in both the etymological and popular senses of the word, by the new horizons I discovered.

Without the encouragement and help of several of my teachers, this work could not have been completed. It is not only for the sake of academic politeness that I must thank especially Pierre Aubenque, Georges Canguilhem, Jean-Toussaint Desanti, and Pierre Hadot. For making the English edition possible, I thank all my American and British colleagues who have passed along their friendly criticisms of the French edition. I thank especially David Balme, who gracefully accepted several criticisms I had directed toward some of his older writings, and Alan Gotthelf, for his effective assistance in bringing this work to the attention of the University of California Press and for his many helpful comments. I also thank the École Pratique des Hautes Études for financial assistance. My debt to Anthony Preus is not measurable, for he has been friend, translator, and advisor, and gave of his time unstintingly to this edition.

ABBREVIATIONS

AELIAN

Nat. An.	*De Naturibus Animalium (On the Characteristics of Animals)*

ARISTOTLE

APo.	*Posterior Analytics*
APr.	*Prior Analytics*
Cat.	*Categories*
De An.	*De Anima*
EE	*Eudemian Ethics*
EN	*Nicomachean Ethics*
GA	*Generation of Animals*
GC	*Generation and Corruption*
HA	*History of Animals*
IA	*Progression of Animals*
Juv.	*Youth and Old Age* (in the *Parva Naturalia*)
Long. Vit.	*Length of Life* (in the *Parva Naturalia*)
MA	*Movement of Animals*
Metaph.	*Metaphysics*
Mete.	*Meteorologica*
PA	*Parts of Animals*
Phys.	*Physics*
PN	*Parva Naturalia*
Pol.	*Politics*
Probl.	*Problemata* (authorship doubtful to spurious)

Resp.	*On Respiration* (in the *Parva Naturalia*)
Sens.	*On Sense and Sensible Objects* (in the *Parva Naturalia*)
Top.	*Topics*

HIPPOCRATES

Fract.	*Fractures*
VM	*On Ancient Medicine*

PLATO

Phdr.	*Phaedrus*
Protag.	*Protagoras*
Rep.	*Republic*
Tm.	*Timaeus*

THEOPHRASTUS

frag.	fragments
HP	*Historia Plantarum (Enquiry into Plants)*
Sens.	*De Sensu et Sensibilibus*

A GREEK-ENGLISH LEXICON

LSJ	Liddell, Scott, and Jones (see Bibliography)

INTRODUCTION

I F ANACHRONISM IS THE CARDI-
nal sin for historians, perhaps historians of
science are the most likely sinners. The very nature of their
subject pushes them almost irresistibly, no matter what they do
to avoid it, toward a conception of scientific discovery that is
evolutionary, if not linear. In fact, even if one has challenged the
fallacious and ultimately ethnocentric concepts of "cultural
progress" and "social progress," one cannot reject the notion of
"scientific progress" so easily. Can this theoretical hurdle be
crossed, now that we have left behind the Cartesian idea that the
sciences developed from an unalterable foundation, and have
substituted for it the idea of the history of science as a succession
of systematic constructs, each of which, in the words of Karl
Popper (1972, 16), "has the character of an approximation
towards [a] new theory"?

This is an ongoing dispute in the history of science. How can
we talk about the development of a science as the heir of the
often metaphysical speculations of ancient authors, who came

1

long before the "epistemological break" that constituted the science in question? The problem becomes all the more serious to the degree that contemporary scientists perceive those authors as their predecessors.

As the present study encounters this problem from the very beginning, perhaps our examination of the status and functions of the classification of animals in Aristotle's writings will contribute to a better understanding of the progress of human efforts toward knowledge. The two views of scientific progress, Cartesian and contemporary, are in competition but usually are not incompatible. They have, in fact, the same root, since they suppose that the theoretical problems that humanity sets itself remain identical throughout the ages. There is here a subtle and biased anachronism: subtle because it recognizes that the material and intellectual means that an epoch uses for resolving these so-called "eternal problems" are relative and variable; biased because it is all too easy for us to forget that *our* problems are not necessarily those of the whole human race.

Nearly all the important literature devoted to Aristotle's "biology" is guilty of this anachronism. But I maintain that it is irrelevant to think of this biology as incomplete; rather, it is radically foreign to us: produced in a world that is gone, it tried to answer questions that we no longer ask. Interpreters evaluate Aristotle's "taxonomic" efforts from a modern point of view. Given that fact, it does not matter much whether they stress Aristotle's inadequacies or his genius, because their dispute is waged in the wrong territory. One might even say that, in this debate, praise is more suspect than criticism, since those who praise him represent the master of the Lyceum as a precursor. And, as Georges Canguilhem remarks in "L'objet de l'histoire des sciences" (1968, 2): "Agreeing to look for, to find, and to celebrate precursors is the clearest symptom of a lack of talent in epistemological criticism. Before joining two sections of road, it is a good idea to be sure first that they belong to the same road."

Historians and commentators who have dealt with Aristotle's classification of animals have most often neglected to address this problem of *method,* in the etymological sense of the word. Clinging to the indubitable fact that Aristotle divided animals into distinct groups, they do not ask whether classifying animals was for Aristotle a theoretical task, as it was to be for the taxonomists of the classical era, nor do they ask what functions the several animal classifications had in Aristotle's system of knowledge — for he presents a variety of orderings of animals, according to different points of view.

Of the large number of interpreters of Aristotle's biology, I shall consider two groups that are chronologically, and especially professionally, distinct. According to the first group, recent historians of science and philosophy, Aristotle tried to construct a natural classification of animals, but was not able to succeed. Questions concerning what route he traversed and how far he had left to go to complete his project are controversial issues among these interpreters. For example, here are three passages from *Histoire de la zoologie* by Georges Petit and Jean Théodoridès (1962, 83, 84, 86):

> Aristotle did not know how to codify his attempts at classification, and it would be illusory to draw from his work an ordered classification; it is substantially there, but is not expressed. . . .

> Relying on the *Historia Animalium* alone, one would be tempted to write that Aristotle, playing with the words *eidos* and *genos*, did not notice the contradictions in giving them different senses from one passage to the next. On the contrary, we think that he was aware of these contradictions or insufficiencies, and was not able to escape them. . . .

> But Aristotle did not substitute anything constructive for the dichotomous classification that he criticized. . . . We think that this decision to take animals genus by genus, following the example of ordinary people who distinguish a class of birds and a class of fish, was in fact a renunciation.

From this point of view, Aristotle could barely have envisaged a rigorous taxonomy, which would have been too ambitious for the period; instead he attempted a less grandiose but more manageable project, leaving the completion of his work to successive generations, who would be better informed. This continuistic notion runs through the interpretation of Léon Robin (1944, 181):

> The exaggerated hopes that Aristotle had for the possibilities of knowledge, as he had conceived it in its essence and method, were overwhelmed by the volume and complexity of the facts that had to be organized. In truth, we should judge that the criticisms raised against this classification are somewhat unjust, considering all the special studies and failed attempts that were required before a satisfactory natural classification could be developed.

If one accepts Karl Popper's idea that scientific theories are constructed by a method of trial and error, one may readily suppose that if a taxonomic effort had been going on for centuries, later naturalists could have profited directly from Aristotle's efforts and failures. As the saying goes, the successors, standing on the shoulders of their great ancestors, could see farther.

Some might object that the book by Petit and Théodoridès, being a survey, is not grounded in an intimate knowledge of the texts, and that although Léon Robin is a specialist on Aristotle, the biological works are not those he knows best. Neither excuse can be made, however, for Pierre Louis, whose *La Découverte de la vie: Aristote* (1975) was published as a kind of conclusion to his edition of Aristotle's biological texts.[1]

In Louis's view, Aristotle found himself faced with the problem of taxonomy as a necessary consequence of his historic initiative, in the etymological sense of *historic:* "He gave to his research on animals a promising new orientation, for he had opened the way to comparative anatomy." If I wished, I could show the danger in masking the formidable problems of historical affiliations with simple but vague expressions like "open the way." But the relationship that Louis sketches here between Aristotelian zoology and comparative anatomy assumes a special importance, to which I shall return later. Louis continues (p. 29):

> That initiative led [Aristotle] to gather more and more information about animals. But it also led him to set himself a difficult problem that he could not avoid raising in discovering comparative anatomy — that of the classification of living things. Classification is for him the result and consequence of the comparisons to which he devotes himself, but it is also the indispensable tool of his research.

Farther on (p. 149), at the beginning of the chapter on "the classification of animals and the scale of beings," Louis is even clearer:

> The comparative method that Aristotle uses for studying anatomy, physiology, and the habits of animals presupposes a rigorous classification of all living beings. However, we find nowhere in the biological treatises a complete table of the various families of animals.

It is not true that Aristotle failed to indicate the direction that a possible classification of animals might take; as Meno does for virtue (Plato, *Meno* 72a), he proposes a swarm of criteria. Without explicit reference, Louis paraphrases a text from the beginning of the *History of Animals* (1.1.487a11 et seq.) in which Aristotle, having enumerated four principles for distinguishing between animals — their mode of life, activities, characters, and parts — goes on to mention ten or more parameters, from habitat to degree of sexual appetite, and finally talks about the distinction between blooded and bloodless and envisages a possible classification according to modes of reproduction and locomotion. Louis concludes from this (p. 151):

> Thus, there are many possible ways to classify animals. Aristotle ultimately chooses, out of this entire range, a way of classifying according to the subject he is studying. . . . But Aristotle doubtless was perfectly aware of the serious drawbacks of an overly simple method. Its principal defect is in depriving the naturalist of a complete and definitive classification.

Aristotle therefore had to have produced, simultaneously with his own biological project, the need for a natural and universal classification. But he is thought to have lacked the means for

constructing such a classification, and to have settled for a diverse multiplicity of partial criteria. And, to crown his misfor-
tune, he could not have avoided being aware, obviously unhap-
pily aware, of his theoretical prerequisite, and therefore his
failure.

Thus, according to Louis, the postulate of the identity of
problems through time leads directly to the identity of the procedures for resolving them. It is historically correct that comparative anatomy, as it was constituted at the beginning of the
nineteenth century, relied on the work of taxonomists of previous generations. One could even say that this was a necessary
sequence, in that one can hardly imagine how it could have
occurred the other way around. Louis transports that necessary
sequence into the past. And, in fact, if one admits that comparative anatomy could, one way or another, have been born several
times, then it must have been the same for taxonomy.

The three interpretations presented above are representative
of the positions that all Aristotelians have adopted on the problem that concerns us, both in works as old as that of Meyer and in
the most recent works.[2] Commentators notice Aristotle's obvious and explicit desire to separate animals into nonarbitrary
groups, and they add a presupposition, so indisputable in their
eyes that they do not formulate it: namely, that Aristotle had to
be trying to achieve that "perfect" classification, the binominal
taxonomy, which we call, perhaps erroneously, Linnaean.[3] So if,
on the one hand, we suppose that Aristotle had a theoretical
project more or less identical with that of Adanson or Linnaeus,
and, on the other, we realize (and it is difficult not to) that
Aristotle did not develop a rigorous binominal classification,
we create from nothing the theoretical problem that confronts
us — that is, the problem of the obstacles that supposedly
prevented Aristotle from achieving the goal that commentators have assigned him. At that point they cease explicating
Aristotle.

It makes little difference whether they then castigate Aristotle's taxonomic weaknesses or try to excuse them: all plead

before the same jury, that of post-Linnaean systematicians, and all plead guilty. At that point the commentators part company, each one proposing his own explanation of Aristotle's classificatory "failure." Let me cite just two of these explanations — first that of W. D. Ross (1949, 115):

> No cut-and-dried classification is to be found in his writings. He is well aware of the difficulties; well aware of the existence of isolated species which fall under no recognized "greatest genus," and of species intermediate between two such genera. But his classification is clear enough in its main lines, and is one which has on the whole stood well the test of time; it was a great advance on anything that preceded it, and no further advance was made before Linnaeus.

Besides the surprising confidence about debates that Aristotle carried on *in pectore,* this passage clearly demonstrates Ross's belief in the continuity of problems through history: if Linnaeus went farther than Aristotle, then they must have been following the same path. According to Ross, Aristotle's effort did not succeed because of the internal difficulty of delimiting genera and species. That explanation is in itself essentially continuist, since this is the same difficulty found by classical taxonomists, and they succeeded just where Aristotle failed.

Another explanation, put forward by Maurice Manquat, Jean-Marie LeBlond, and Pierre Louis, emphasizes an external obstacle to the practice of classification as such. This epistemological obstacle is, they say, that of ordinary language: in not breaking with ordinary language, Aristotle could not develop a rigorous classification. Indeed, what scientific innovation has not run up against the difficulty of expressing its contribution in the language inherited from previous generations? But Aristotle seems in a way to refuse to engage in that struggle. Thus, LeBlond (1945, 168, n. 94), commenting on the passage in *Parts of Animals* (1.2.642b15) in which Aristotle notes that certain groups, although "natural," do not have names, writes:

> Aristotle did not seek to fill this gap in popular terminology, and he did not create names. That is surely one of the reasons for his failure in natural classification; he had no idea of rational terminology, a

necessary tool of classification, which would be the glory of Linnaeus.

This lack of linguistic daring on Aristotle's part is all the more surprising because he himself declares that existing language is often an inadequate tool and one should not hesitate to create terms when the need arises.[4] Doubtless this explains why Louis (1975, 156), unable to account for Aristotle's supposed linguistic conservatism, claims that it was common practice in Aristotle's day:

> This respect for contemporary language is not, in any case, limited to Aristotle. All ancient naturalists share it; unlike modern scientists, they did not use any scientific nomenclature for designating plants and animals.

In the context of LeBlond's general interpretation of Aristotle, the taxonomic failure is no small matter. If a rigorous classification is necessary for a coherent "biology," and if it is really in the study of living things that Aristotle forged most of his philosophical concepts (which is what LeBlond thinks), then the failure in classification is fundamentally the failure of Aristotle's philosophy in general.

> It is incontestable . . . that Aristotle's logic and his whole philosophy were inspired in large part by a classificatory ideal: syllogism and definition, in theory as in practice, are determined by the notions of genus and species, which are decidedly biological in character and do not find their perfect application other than in the realm of living things. In biology proper, however, the classificatory effort appears not only less successful than the explanatory effort, but also less carefully thought out. Even the theory is developed less confidently.   *(LeBlond 1945, 59, n. 3)*

Finally, we can reproach Pierre Louis, who knows the texts very well, for not having taken his own discoveries seriously enough. As a matter of fact, in an article (Louis 1956) written long before the work cited above, he lays out all the procedures that Aristotle used for not getting too far away from the vernacular. That vernacular, as Manquat properly emphasizes, seems to us scientific only by a retrospective illusion.[5] Louis gives examples of periphrastic descriptions, uses of the participle *kaloumenos* ("so-called") with less-used names, and so on. Aristotle intentionally spoke the language of his informers — travelers, hunters, fishermen, farmers. If Aristotle was careful to stay close to ordinary language in his discussions of animals, no doubt it was primarily (as Louis suggests) because he wanted his hearer (or reader) to be able to immediately identify the animals named. But that Aristotle would sacrifice his scientific project itself to the goal of popularization is a truly incredible idea. As a matter of fact, Aristotle does resort to neologisms, whether invented by him or by others, when necessary: for example, the noun *entoma,* "insects," which is etymologically clear, and the term *selache,* "selachians," which Pliny the Elder (9.40) assures us Aristotle created.[6] And Aristotle clearly innovated the fundamental division between blooded and bloodless animals.[7]

We can grant to Louis and LeBlond that Aristotle could not achieve a rigorous taxonomy, in the modern sense, without breaking with contemporary language. But we can avoid falling into their aporias by simply supposing that, since Aristotle did not break with the linguistic usage of his time, he did not have any such taxonomic project. David Balme (1975, 188), correctly observes: "The belief that there must be a classification in the background rests on the assumption that Aristotle, like every good pre-evolutionary zoologist, put systematics first in zoology and morphology first in systematics." A critique of the linguistic explanation of Aristotle's supposed taxonomic failure could take us very far afield. To adequately appreciate the force of the argument presented by LeBlond and Louis, we would first have to ask, "What is the theoretical function of naming?" Let us simply say that for the Greeks in general, and for Aristotle in particular (as we shall see later), naming cannot have an epistemic function, since naming is not defining. In fact, name-giving has not had a really theoretical function in natural science except during a rather brief period, approximately the last two-thirds of the eighteenth century, and this is well known to historians of science. To suppose that Aristotle intended such a function is mistaken — the Greeks would say *atopon:* without significance, because uprooted from the place in which it can be significant.[8]

Let us turn now to a second group of commentators, very different from the modern Aristotelian scholars we have been discussing, but whose conception of his taxonomic project is no less instructive. These are not so much the creators as the heirs of the triumph of animal systematics — those who received a completed taxonomy from the hands of their predecessors and turned to look at new problems.[9] The most representative member of this group is doubtless Georges Cuvier, the last great fixist, but nevertheless the founder of biology, as I shall explain later. (By "fixists" I mean those classical taxonomists who believed that the kinds of living things are denumerable and permanent.) The common practice of his predecessors, even of the greatest, such as Buffon, was to dismiss the history of their science with mere summary doxographies. Cuvier, by contrast, was a true historian of science, and this part of his work influenced his own original contributions. Thus, we should look closely at the two lectures that he devotes to Aristotle in his *Histoire des sciences naturelles.*[10] Here are several passages:

> We should consider Aristotle one of the greatest observers who ever lived; but without any doubt his genius for classifying was the most extraordinary ever produced by nature.   *(Vol. 1, p. 133)*

> The *History of Animals* is not exactly a treatise on zoology, that is, a set of descriptions of various animals; rather, it is a kind of general anatomy, in which the author treats the general organizational features that various animals present, and in which he describes their differences and similarities, relying on a comparative examination of their organs, and in which he lays the foundation of the great classifications with the most perfect exactitude.   *(Vol. 1, p. 147)*

> However, since Aristotle did not think it necessary to draw up a zoological chart, some people have supposed that his work lacked method. Assuredly, those people have only a very superficial understanding.   *(Ibid.)*

Aristotle, right from the beginning, also presents a zoological classification that has left very little to do for the centuries after him. His great divisions and subdivisions of the animal kingdom are astonishingly precise, and have almost all resisted subsequent additions by science. *(Vol. 1, p. 148)*

Notice that Aristotle's groups are formed in a very natural way, and that only their disposition leaves room for criticism.
(Vol. 1, p. 149)

This is a surprisingly enthusiastic appreciation from that scrupulous and often critical spirit. Georges Pouchet (1884, 353) finds in it "an almost suspect exaggeration." Doubtless, Cuvier falls into the anachronism that we denounced above, seeing in Aristotle a systematician in the modern sense. But it is astonishing to find Cuvier regarding Aristotle as the author of a rational classification, an appreciation far in excess of the most indulgent emanating from recent scholars. As I shall show, a precise understanding of Cuvier's judgment of Aristotle led me to an interpretation of Aristotle's biological project that will perhaps be more accurate and more precise than those developed heretofore.

This is because, in a way, Cuvier was right. The classifications of animals proposed by Aristotle are not ordered subjectively, and they do have an unarguable systematic value. Thus, for example, the distinction between blooded and bloodless animals, even if displaced by the distinction between vertebrates and invertebrates, remains one of the great points of articulation in modern classifications.[11] For another example, the distinction between animals based on the degree of development of their offspring, proposed in *Generation of Animals* 2.1, shows an embryology that is more advanced — in the sense that it is less metaphysical — than that of the partisans of preformation and epigenesis. Ernest Haeckel (1900, 54) was right, then, in saying of Aristotle's embryological ideas that "it was not until our own day that many of them were fully appreciated, and, indeed we may say, discovered afresh." And the anthropocentric classifications that prevailed from Theophrastus to the seventeenth century give a negative measure of the objective integrity of Aristotle's distinctions.

It is not, in any case, a matter of indifference that the praises cited above flow from the pen of Cuvier, the incontestable father of comparative anatomy, that is, a biologist who had taxonomy (natural history) behind him and who made of it a tool and not the end of his research. Aristotle seems to use his classifications in the same way, without appearing to look for them for their own sake. Thus, although every construction of a classification is at least partly inductive, Aristotle more willingly leaves vague the distinctions that are most immediate, and consequently easiest to establish, than the more general, which he ought to have derived from the first. For example, he does not distinguish reptiles, a class that he neither isolates nor names, from the batrachians, a class that he also does not name.[12] If one sticks to the texts, one would have to agree with Balme (1975), who gives a section of his article the heading "Evidence That Aristotle Did Not Classify Sub-Genera." That states a fact that is impossible to understand if one thinks Aristotle had a taxonomic project. We shall meet many facts of this kind in succeeding pages.

Indeed, one might be led to believe that Aristotle had developed a relatively rigorous classificatory method, but did not use it for classificatory ends. We shall begin by studying this classificatory procedure, which utilizes, on the one hand, the logical instrument of division *(diairesis),* which Aristotle borrowed from the Academy, and on the other, the three concepts of *genos, eidos,* and *diafora,* which an entire tradition has read as "genus," "species," and "specific difference." We can then try to see the status and functions that animal classifications had for Aristotle.

1

DIVISION AND DEFINITION IN ARISTOTLE
THE MEANING AND THE LIMITS OF THE CRITIQUE OF PLATONISM

ALTHOUGH PLATO WAS INTER-
ested in zoology, we cannot, if we re-
strict ourselves to his extant texts, consider him a real predecessor of Aristotle in this area. Certainly, there are some signs of
zoological research at the Academy under the direction of its
founder, notably exercises in division, to which I shall return
later; but between the zoology and anatomy of the *Timaeus* (the
zoology is much less developed than the anatomy) and the biological treatises of Aristotle there is a considerable difference in
kind. The very quantity of Aristotle's observations on living
things allows us to consider him as, in a sense, the founder of
zoology, despite the loss of texts by, for example, Empedocles,
Democritus, or Anaxagoras.[1] To be sure, Aristotle placed zoology in a larger frame constructed from metaphysical principles
ultimately related to those of Plato: the study of animals is part
of physics; it is a theoretical science (in the sense of *episteme*) of
the real in becoming. And I, like almost all interpreters of Aristotle's biological texts, must assert that his zoology is essentially
metaphysical, in that it depends fundamentally—at least in

13

theory — upon metaphysical principles and has for its principal function the corroboration of those metaphysical principles by illustrating them. But the internal coherence of Aristotelian zoological research seems *to us* to give it an autonomous status, which gives the biological corpus a properly scientific posterity independent of its metaphysical framework. In any case, later naturalists have often taken Aristotle as a predecessor, which they never have done in the case of Plato.

Nevertheless, there is a domain of zoological research in which, at first glance, Aristotle may seem to have retrogressed in comparison with Plato and his disciples: namely, in respect to the development of theoretical tools that could have permitted an objective classification of animals. All scholars of Aristotle know that he rejected the Platonic method of "division" *(diairesis)*, criticizing it, often harshly, in many places, notably in the *Analytics* and *Metaphysics*. What interests us here is that he also rejected it as a means of dividing animals into relevant groups in that genuine "discourse on biological method" which is the first book of the *Parts of Animals,* where a long and detailed condemnation occupies Chapters 2 and 3. But every classificatory enterprise sooner or later relies on "divisions"; hence, one might suppose that the break with the Academy on this point was costly for Aristotle. Even if none of Plato's works was devoted entirely to zoology, nevertheless in the *Phaedrus* (266c), for example, Plato defines a method, which he says belongs to the dialecticians, that might well have led to important taxonomic developments. In one passage (265c9 – 266c1), Socrates persuades Phaedrus to define things by a double movement of division *(diairesis)* and collection *(synagoge);* this method allows the dialectician to take account of the articulations that exist in reality. When Socrates adds, "Believe me, Phaedrus, I am myself a lover of these divisions and collections, that I may gain the power to speak and to think, and whenever I deem another man able to discern an objective unity and plurality, 'I follow in his footsteps where he leadeth as a god' " (266b3, trans. Hackforth), Plato seems, through the mouth of his master, to invite his own disciples to apply this method to new domains. The method of "division" developed in the *Sophist* and *Statesman* is even better known, and it is this version of Platonic division, doubtless later than the other, that Aristotle seems to have as his target in his critique both in the *Parts of Animals* and in the logical and metaphysical texts. And despite the loss of almost all the post-Platonic Academic texts, there is evidence that division held a prominent place in the teaching of the Academy even during the lifetime of Plato.[2] It is, moreover, difficult to imagine that Aristotle would have attacked so bitterly a method that its own authors considered obsolete.[3]

Yet zoology, and especially animal systematics, seems to offer itself as a privileged field for this method. We might well wonder whether it was not Aristotle's abandonment of the method of division that closed off his access to a rigorous taxonomy. This appears all the truer because, until recently, most commentators have agreed that Aristotle replaced division with the syllogism. And it is incontestable that in zoology, as elsewhere, Aristotle thought that he must use syllogisms.[4]

We must therefore evaluate more precisely the zoological consequences of Aristotle's abandonment of the Platonic method of division. In his logico-metaphysical writings, Aristotle rejected division as a means of demonstration. It would seem that this rejection of division should have extended to all domains, including zoology; and one might discover, especially when considering the first book of the *Parts of Animals,* that the zoological rejection of division was nothing more than a specific application of a more general attack on Platonism. In that case, an extra-zoological debate would have caused Aristotle to abandon the Platonic project of constructing divisions. One might wonder whether, whatever advantage Aristotle may have gained by abandoning *diairesis,* he may not have sacrificed the construction of a rigorous classification to this theoretical advantage.

Since, on the other hand, Aristotle could not carry out his study without a minimum of organization of the immediate data of zoology, he would have had to fall back on the empirical classifications of common sense. That is how interpreters generally understand his statements when, after a long and hard critique of the method of division in the *Parts of Animals,* he indicates several times that it is preferable to retain traditional classifications:

> The proper course is to endeavor to take the animals according to their groups, following the lead of the bulk of mankind, who have marked off the group of Birds and the group of Fishes.[5]

This conception of Aristotle as retrogressive in classification in comparison with Plato tends in the same direction as that of Jean-Marie LeBlond. In fact, for LeBlond, although classification "governs . . . the theory of demonstration" (1939a, 295), nevertheless "Aristotle does not appear to us to have been primarily a great classifier" (1939a, 298); so that "choosing the great genera on the basis of common sense . . . agrees well with the conformist attitude that was rather habitual for Aristotle" (1945, 176, n. 116).

Pierre Louis disagrees with this interpretation.[6] For Louis, the Aristotelian criticism of Platonic division, when it is applied to zoology, is aimed at the taxonomic weakness of the Platonic method. Far from resigning himself to a retreat in relation to the Academy on the problem of taxonomy, Aristotle (Louis argues) blamed the Platonists for their not having found a satisfactory solution. In fact, commentators on the passage of the *Parts of Animals* in which Aristotle criticizes Plato's method of division generally find three main objections: the divisions split up natural groups; in using opposed differences, they divide according to privations (e.g., winged versus wingless); and they lead to useless repetitions.[7] It matters little whether this summary is faithful or not to the content of this famous passage, whose difficulties and corrupt text always surprise anyone who looks at it closely.[8] What interests us here is that by this criticism Aristotle seems to move toward the construction of a rigorous taxonomy, if only in saying what it should *not* be. This is what Louis thinks when he writes (1955a, 300):

> These criticisms . . . demonstrate that Aristotle understood very well, when he composed the *Parts of Animals,* the conditions and characteristics of a perfect classification.
>
> But he was not in a position to attain this perfection. Thus, he was satisfied with a provisional classification that was only a simple working tool, like the imperfect hammer that serves for the forging of a better, as Spinoza says. He simply used the ordinary classifications admitted by common sense, the distinctions that everyone has the habit of establishing.

Thus, in Louis's view, Aristotle's theoretical path followed these steps: (1) Educated in the Academy, Aristotle understood that the taxonomic tool handed down by his master was imperfect, and he criticized it severely; that is, in the long passage devoted to it in the *Parts of Animals,* he does not recognize any positive function for it. (2) Aristotle attempted to reform the Platonic taxonomy. At this point, the theoretical advance would already have been considerable: Aristotle would have grasped, at least negatively, some of the possibility-conditions of a rigorous taxonomy. But (3) he failed, for reasons Louis does not give us, to construct an adequate procedure, and he resigned himself to traditional empirical distinctions. Still, that does not represent a retreat in comparison with Plato, for an approximate classification that one knows to be approximate is better than a classificatory process that one believes is rigorous but is not: criticism is often the first footing for a new foundation. Finally, we should note that if Louis were right, Aristotle would have at last acted like a good disciple of his master in pushing forward a reformation of the Platonic classificatory method. Thus, in this sense, Louis's interpretation is actually the inverse of LeBlond's: Aristotle did not break with Platonism, but continued the task of his master with or without a clear conscience.

One could easily criticize Louis's position by showing that it is a reconstruction of a theoretical itinerary for Aristotle that has no real textual basis. For Aristotle never represents "common" divisions as provisional, second-best, and theoretically insufficient, which he intended eventually to supersede. Not that he accepted traditional divisions without criticizing them, but his criticism did not go in the direction one would expect from Louis's interpretation. In fact, in the same section of the *Parts of Animals,* having said a few lines earlier that the traditional classification is correct (*orthos,* 4.644a16), Aristotle writes:

> Perhaps, then, the right course is to speak of some affections in common by genera, wherever the genera have been satisfactorily marked off by popular usage.   *(644b1, trans. Balme)*

He thus hints that "people" (*anthropoi,* 644b3) can be wrong and that the natural philosopher should be careful not to repeat their errors under his own name. There is here a criticism of the *content* of the traditional divisions, but not of their *method,* which Aristotle considers right, sufficient, and definitively established.[9] Finally, no text exists that would support the belief that, after his criticism of Plato, Aristotle had any intention of developing a new method of classification. But I shall soon develop a more radical criticism of Louis's notion of a reforming Aristotle who tried to continue the taxonomic efforts of the Academy.

Finally, let us examine an interpretation that goes farther than the others, that developed by G. E. R. Lloyd in a rich and subtle article, "The Development of Aristotle's Theory of the Classification of Animals" (1961). I should say at the outset that, in a way, this interpretation is a synthesis of the two already discussed, in that it presents Aristotle as at first following the Platonic way, then breaking with Plato in order to develop a method of his own.

Although Lloyd's study begins with a statement that Jaeger's interpretations are outdated, it seems to me deeply Jaegerian in spirit. Lloyd actually tries to discern, through texts arranged chronologically, the steps of an evolution that goes from a more or less Platonic attitude in regard to division, to an entirely original mode of classification of animals. In the works not devoted specifically to biology, Lloyd sees three successive stages in Aristotle's appreciation of the taxonomic powers of division: in the *Topics* and *Categories,* one finds "a somewhat uncritical acceptance of the method of division" (p. 69); in the *Analytics* and *Metaphysics,* "there is the analysis of division which led Aristotle to reject it as a proof, but to retain it as a useful method for definitions" (p. 70); finally, relying on a single text in the *Politics* (4.4.1290b25 et seq.), Lloyd thinks he can extract a last stage, in which "there is the substitution, for the method of division, of a method of classifying species according to the combination of the varieties of their necessary parts" (p. 70).

Turning next to the biological treatises of the Aristotelian corpus, Lloyd believes that, by a close and acute analysis of the texts, he has found three stages there, too. In the first book of the *History of Animals,* under a strong Platonic influence, Aristotle "may . . . have attempted a classification of animals into contrary groups" (p. 79); and Lloyd believes that he can discern in this text a struggle in Aristotle between dichotomic and anti-dichotomic tendencies: "It seems, then, that we have evidence here, of Aristotle correcting himself, and the first statement may represent his ideas at a stage when he had not entirely rejected such Platonic dichotomies" (p. 71). The second step would be the criticism of Platonic division in the first book of the *Parts of Animals* (chaps. 2 and 3). The last step would be taken in *Generation of Animals* (2.1), "when a classification by the organs of locomotion is rejected, and a new test (by degrees of perfection of the offspring) . . . is introduced" (p. 79).

Parts of Animals 1.2 – 3 would thus be a purely critical passage, marking the definitive abandonment by Aristotle of the method of division before he began to elaborate his own non-diairetical process of classification. This idea leads Lloyd to conclude, among other things, that this passage "represents a later stage of development than any of the passages from the *Organon* or the *Metaphysics* considered above" (p. 71).

In considering Lloyd's interpretation as a whole, one should keep in mind the unreliability of any analysis that rests on a chronology of Aristotle's extant works. One could also criticize the circular reasoning to which this method leads Lloyd — for, to take just one example, he bases his claims about the evolution of Aristotle's thought concerning division, and thus classification, on the later date of *Politics* 4 in relation to *Metaphysics* Zeta, but, on the other hand, one of the most solid arguments for this temporal relationship is that the *Politics* reflects a later stage in Aristotle's theory of classification. This is one of those dilemmas that stymie historians of philosophy: although they know that the thought of every philosopher evolves in the course of his life — and in Aristotle's case the study of that evolution has been one of the most fertile domains for recent interpreters — they should make the least possible appeal to that evolution in trying to resolve the contradictions they believe they find in the texts. For the evolutionary explanation is fundamentally tautological, since it ultimately says that the positions are different because they are not (any longer) the same, and thus it is all too often a hermeneutical band-aid.

Thus, if we look at Lloyd's interpretation in detail, we can easily show that his reconstruction does not accord very well with the texts. Let us look successively at each of the three stages that Lloyd traces in the Aristotelian notion of the classification of animals in the biological corpus: *History of Animals* 1.1–6; *Parts of Animals* 1.2–3; and *Generation of Animals* 2.1.

Lloyd thinks he has found in the first of these texts a method of animal classification that is very close to Platonic dichotomy. In the beginning chapters of the *History of Animals,* Aristotle first divides parts *(moria)* into *homoiomeries* and *anhomoiomeries. Homoiomeries* are homogeneous, composed of parts called by the same name, as flesh and bone: "thus fleshes are divided into fleshes" (1.1.486a6). *Anhomoiomeries* are composite and divide ultimately into *homoiomeries;* thus the hand is divided not into hands, but into flesh, bone, and so on.[10] Then he arranges animals according to four criteria: mode of life, activities, character, and parts (1.1.487a10), introducing for each criterion a great number of distinctions. A rapid reading of these distinctions could lead one to think, like Lloyd, that Aristotle tended to approach the Platonic dichotomous method. That is also the opinion of Pierre Louis, who, in his edition of the *History of Animals,* sees in 1.1.487a14 ("Some are water-animals, others land-animals. There are two ways *(dichōs)* of being water-animals" [trans. Peck]) "a reminder of the method of division in the *Sophist* and *Statesman.*" An attentive reading leads, in my opinion, to a totally opposite conclusion: this entire passage is a criticism of dichotomy, not methodological like that of *Parts of Animals* 1, but what might be called "practical." In fact, Aristotle seems to be struggling, in all the distinctions he makes, to show that the binary divisions of the Platonists are unable to account adequately for the diversity of reality. Thus, if we go on from the passage quoted above:

> There are two ways of being water-animals. Some both live and feed in the water, take in water and emit it, and are unable to live if deprived of it: this is the case with many of the fishes. Others feed and live in the water; but what they take in is air, not water; and they breed away from the water. *(Trans. Peck)*

There might seem to be here a dichotomous division, in the Platonic manner, of the class of aquatic animals into animals with gills and animals with lungs. Farther on in the *History of Animals* (8.2.589b13), Aristotle clarifies his thinking about these two organic arrangements by showing that they are two different ways of arriving at one and the same goal:

> For the fact is, some aquatic animals take in water and discharge it again, for the same reason that leads air-breathing animals to inhale air: in other words, with the object of cooling the blood. *(Trans. Thompson)*

Nevertheless, a few lines after the text cited from 1.1, Aristotle adds:

> Some animals, although they get their food in the water and cannot live away from it, still take in neither water nor air: examples are the sea-anemones and shellfish. *(1.1.487a23, trans. Peck)*

In effect, then, the supposed dichotomous division posited above leaves a "remainder" that, from a classificatory point of view, takes from it all theoretical validity.

All the examples that follow confirm this interpretation. Rather than enumerating individual cases, it would be interesting to give a list — if not exhaustive, at least long enough — of the procedures used by Aristotle in this passage that give evidence of the inability of dichotomy to account for reality.

The procedure we have just seen is that of dividing a family in two and then indicating that there are members of this family that are not in either of the two subfamilies.

Sometimes Aristotle divides from the outset into more than two classes, and of course it is well known that Platonic division is binary.[11] Thus, terrestrial animals either walk, crawl, or undulate (487b20). This tripartition, which includes the distinction between the reptilian movement of snakes and the undulation of worms, shows even in its form the anti-dichotomous position of Aristotle. For in this instance it would have been easy for him to respect the Platonic procedure by making two binary divisions: first, one between animals that move with feet and those that move without feet; and then, among the latter, a distinction between those that crawl and those that undulate. Thus, Aristotle did not remain faithful to Academic orthodoxy even when he could easily have done so.

Sometimes Aristotle shows that the subgroups that divide the group under consideration overlap. Thus, if one can divide animals into gregarious and solitary, there are some that are in both categories (1.1.487b34); for example, "man is in both groups" (488a7).[12]

Aristotle could also elucidate the relative and precarious character of the distinction being made. Thus, animals are divided into wild and tame, but some can pass from one category to the other; and, in any case, as Aristotle repeats several times, "all tame animals are also found in a wild condition."[13]

There is therefore no reason to accept Lloyd's argument that the beginning of the *History of Animals* adopts the Platonic

method of division, even approximately, and that Aristotle condemns it only in *Parts of Animals* 1. The only difference between
the two texts on this point follows from the difference between
their functions: one is a theoretical exposé, explicitly aiming at a
condemnation of the Platonic method, and the other is a preliminary ordering of zoological data. But the *History of Animals*
seems to take for granted the rejection of the Academic method.
The two texts are therefore consonant. Did the empirical classifications at the beginning of the *History of Animals* motivate
Aristotle's development of a radical theoretical critique of the
method of division, or, on the contrary, was the theoretical
critique developed first, and only applied in the *History of Animals?* The content of these texts, at least on the point that concerns us now, does not permit us to decide this question, but that
does not matter much for our study.[14]

This anti-dichotomous reading of the first distinctions made
in the *History of Animals* does not, however, prevent them from
having a preparatory and provisional character — one, in other
words, that can be surpassed. Thus, in the case of the division of
animals into terrestrial and aquatic, Aristotle will later feel the
need to "define further" what one should understand by
"aquatic," in distinguishing those that absorb water for their
cooling and those that do it for their food.[15]

We must now return to a closer look at what Lloyd takes to be
the second stage, chronologically, of Aristotle's position in regard to Platonic division, the condemnation of this method
presented in *Parts of Animals* 1. It seems that interpreters have
paid attention to the *content* of Aristotle's criticism — there is no
harm in that — but have too often neglected to tell us the precise
target of the criticism.

Sticking to the letter of the text, we see that in the critical
passage in question, Aristotle never attacks *division,* but rather
dichotomy, a perverse form of division.[16] Dichotomy is a species
of division that is characterized by dividing into two, as the
etymology indicates, and into two opposed groups, as Aristotle

makes clear in the expressions cited in note 16; this is also clearly apparent in the Platonic texts. I say "opposed" and not "contrary," and we must, in a later chapter, determine the difference that Aristotle finds between the two. On this point, David Balme (1975, 185) recognizes his own past errors and remedies them:

> Past interpreters, including myself, have assumed that *PA* 1 criticises the general procedure of division as given in the *Organon* and *Metaphysics,* but in fact its target is only "dichotomy," by which Aristotle means division into two classes characterised by a single differentia and its opposite.

And in fact dichotomy, in its formal rigidity, deserves the reproaches enumerated above. In the first place, it splits up natural groups: on the one hand, by making division according to more than *one* difference impossible (644a8), thus putting together heterogeneous animals; on the other hand, by positing that every animal is either aquatic or non-aquatic, thus leading us to divide an obviously homogeneous group like that of the birds. In the second place, this procedure divides by privation (e.g., winged/wingless) and by accidents (643b23: first division, wingless/winged; second division, winged into domestic and wild). Finally, it implies repetitiveness, since it will be necessary, in order to get to the individuals, to traverse the whole chosen line from which branch unretained segments in each division, as can be seen in Plato's *Sophist* and *Statesman:* for example, after the division footed/footless, we have the division biped/non-biped, although "footed" is already contained in "biped" (642b7).

What about the relationships between this section of the *Parts of Animals* and the chapters of the *Analytics* and *Metaphysics* that criticize division? In fact, despite what Lloyd thinks, it is only by homonymy that one might see a convergence between these texts: both condemn division. But there is a double equivocation here: on the one hand, the logical and metaphysical texts do indeed condemn the illegitimate apodictic pretensions of division; but to say that division does not *demonstrate* in no way refuses it a place in the search for truth. During the last several years, we have seen a kind of rehabilitation of Aristotelian division, although previous interpreters, fogged in by syllogistics, had the tendency to consider division as a kind of fossil concept, a vestige of an obsolete form of Aristotle's thought. I have argued elsewhere that Aristotle gives to division a fundamental place in his cognitive strategy, that of constructing definitions.[17] On the other hand, *Parts of Animals* 1, far from condemning division, seems to me, if one reads the text correctly, rather to recommend its use.

On this point, too, Balme has put the Aristotelian text into its true perspective. All interpreters before Balme saw in *Parts of Animals* 1.2–4 a criticism of Platonic division — which is, in fact, as we have just seen, not a criticism of division in general but a rejection of dichotomy alone — an attempt on Aristotle's part to develop a method of classification that would be his own, or (in their way of seeing things), a non-diairetical method. Thus, Louis and LeBlond, in their respective translations, give as the title of *Parts of Animals* 1.4, "Principles of a Rational Classification," or more simply, "Principles of Classification." From this point, opinions diverge on the question of whether or not Aristotle succeeded in establishing his own method: Louis, as we have seen, does not think so, while Lloyd sees in the text of the *Politics* to which he refers — and I, too, will be led to examine it in more detail below — a first sketch of this new non-diairetical Aristotelian method of classifying animals.

But the need for finding in Aristotle a "new" method, that is, a non-diairetical method, of classifying animals is not a consequence of what we find in the text, but of the position of the commentators themselves: since they believed that Aristotle abandoned division, understood in the most general sense of the word, it would indeed have been necessary that his method, to forestall serious contradiction, be non-diairetical. But the text not only does not support this interpretation, it actually says the opposite. Not only is it indubitable that Aristotle's criticism of Academic method is not put forward in order to support the construction of his own method, but there are also two bits of textual data that are incontrovertible. In the first place, as we have seen, it is dichotomy, and not all division, that Aristotle condemns. Second, at the end of *Parts of Animals* 1.4, when he recapitulates, as he habitually does, the route traveled, Aristotle writes:

> As for division, we have said in what way it is possible to resort to it usefully, and why dichotomy is both impotent and empty.
>
> *(644b17)*

Thus, far from throwing out the diairetical baby with the dichotomous bathwater, Aristotle maintains that there is a fertile use of *diairesis:* to put it another way, Aristotle substitutes a non-dichotomous *diairesis* for a dichotomous *diairesis.* That is why, after attacking Düring and Lloyd for adopting the position I have been criticizing, Balme writes, absolutely correctly (1972, 105):

> The method that [Aristotle] proposes instead [of dichotomy] is itself another form of division. It seems more likely, therefore, that his purpose here is to apply the logical technique of division to zoology, and to show that it must be conducted by multiple differentiae if it is to work.
> Aristotle's aim in using division, again, does not seem to be classification, but definition.

We must therefore say something about Aristotle's diairetical method in zoology.

We will see in detail in a later chapter how the Aristotelian process of division permits the search for specific differences. But we can try here to locate Aristotle's diairetical method more exactly in relation to dichotomy. We will be helped by another passage from the *Parts of Animals* 1 (3.643b9), which I have divided into numbered sections to facilitate our examination of the several issues arising here.

1. In general, this is the necessary result of dividing by such a single difference.[18]
2. But we should try to take animals according to the kinds *(kata gene)*, allowing ourselves to be guided by ordinary people, who define *(diorisantes)* the kind "bird" and the kind "fish."
3. Each of these kinds is defined by a great number of differences, which dichotomy is unable to do.
4. With dichotomy, in fact, either one cannot obtain them at all [the animals according to their kind], for one animal will fall into several divisions *(diaireseis),* and also contraries will be found in the same [division];
5. or there will be only one difference, and this either as simple or as compounded will be the final species *(teleutaion eidos).*
6. If one does not obtain the difference of the difference, one can only make the division continuous, in the way one might unify a speech, by conjunction *(syndesmo).*

Phrase 1 is a transitional summary, since Aristotle has just enumerated the problems with dichotomy. The periphrastic "dividing by such a single difference" obviously refers to dichotomy and not to all division. I shall speak again in detail about the "specific difference" when I examine the terms *genos* and *eidos,* but now, too, we must ascertain a bit of the content of the Aristotelian critique. Aristotle explains his position in several places, and notably in the passage that follows immediately (643b19 – 24), which Balme translates thus:

I mean the sort of thing that comes about if one divides off the *featherless* and the *feathered,* and among the *feathered* the *tame* and the *wild* or the *white* and the *black;* for *tame* or *white* is not a differentiation of *feathered* but begins another differentiation and is accidental here. This is why one should divide off the one kind straight away by many differentiae, in the way that we say.

Aristotle is arguing here that dichotomy has gotten itself into a dilemma: either it remains at the interior of a homogeneous domain, namely, the domain that is designated by the expression "single difference," and after having divided the animals into winged and non-winged, it will divide the winged into split-winged and full-winged, and so on, and end up by defining an animal simply by the conformation of its wings; or it will break the homogeneity of the domain in which it is operating, by dividing animals into winged and non-winged, and then winged into wild and tame. The latter case has the result of dividing according to an accident, for, whether or not it has wings, it is accidental that an animal is wild or tame. There will then be a need for several divisions according to several differences in order to characterize an animal: in one division — it would be better to say in English "one axis of division," that is, a domain in which one makes several divisions in reference to the same object, thus specifying new divisions in relation to others as the division proceeds — we would give the shape of the wings, in another axis a division according to the mode of life, and so on. We notice immediately that the method anticipated by Aristotle is indeed a method of division.

In phrase 2, Aristotle states his preference for the common method over the dichotomous method: this position has already been noted; we shall look at the real reason for this preference below.[19] But first we must note that common sense is deemed superior because it *defines* the animal families it distinguishes; that is, in the etymological sense of the word *diorizein,* it draws a boundary around them. This boundary is, as phrase 3 says, *polygonal* in the proper sense of the word: "each of these kinds is defined by a great number of differences"; each animal family is bounded on several sides, according to what I have called "axes of division." It is this ability to define that we must retain from common sense. And the expression that comes immediately after should not be understood (as it is by all other interpreters) as a reaffirmation by Aristotle that we should not use dichotomy — such a repetition would surely be superfluous — but rather as an affirmation that dichotomy is unable to close in on reality from several sides.[20]

Ultimately, it is the one-dimensional character of dichotomy that is its fundamental vice, from which flow all the others. And it seems to me that in the light of the criticisms that phrases 4 and 5 direct against dichotomy (not new and only repeated here), there appears an idea of great classificatory importance.

These criticisms are the following: first, with the dichotomous method, the same animal falls into several divisions. Aristotle had given an example earlier: "Among the octopi, some are put with the land animals, others with the water animals."[21] Another defect of dichotomy is that it puts heterogeneous entities in the same division; and by "heterogeneous entities" one must here understand animals that are specifically different but that the division would inappropriately bring together by virtue of one of their characteristics; thus, the dichotomists put both man and bird in the class of bipeds, although "being biped is other and specifically different" for them (643a3). Phrase 5 takes up again the criticism of the one-sidedness of dichotomy, about which I have already said enough.

Considering only the first criticism that Aristotle makes of the Platonic method, we should ask ourselves into which category he puts the octopus, aquatic or terrestrial (swimming or walking)? The difficulty here is logical far more than zoological, for the dichotomist must decide whether the octopus is aquatic or terrestrial (a swimmer or a walker) under pain of not being able to go any farther with his division. No such problem exists in the Aristotelian diairetical method, which we are calling "polygonal," for upon examination we find that the other properties of the octopus do not depend on its means of locomotion.

Perhaps there is here an initial outline of the idea that among what I have called "axes of division" some are more relevant than others. In contrast to Platonic division, the Aristotelian method can develop a hierarchy of the determinations of living things. This separation between what is essential and what is not essential in animals permits Aristotle to make distinctions that possess what I called in the Introduction "objective coherence": the most immediately striking characteristic of the animal — for example, the ambiguity between swimming and walking for the octopus — is not necessarily the most significant. Thus, Aristotle makes no allusion to this property when he compares the octopus to other cephalopods — for example, in the *History of Animals* 4.1.523b21 et seq.

Phrase 6 poses a knottier problem of interpretation. Although there is no doubt that Aristotle criticizes the Platonists for not being able to draw correctly "the difference of the difference," it is difficult to decide whether recourse to connection by conjunction is a second-best alternative that the Platonists were forced to use despite themselves, or if it is a positive recommendation on Aristotle's part. But the first difficulty arises from the very expression "obtain the difference of the difference." In fact, just as for division in general there are both good and bad ways of dividing, so too there are good and bad ways of taking "the difference of a difference." The bad way, obviously that of the Platonists, involves the intersection of what we have called two axes of division. Immediately after the passage we are discussing, Aristotle says: "I am talking about what happens to those who divide animals into wingless and winged, and the winged into tame and wild" (643b19). We have seen that the intersection of the axis "form of the wings" and the axis "mode of life" had to be condemned because it leads to division according to accidental attributes. But there is a good way to "obtain the difference of a difference," as is indicated in the *Metaphysics* (Zeta 12.1038a9), where this expression signifies continuing the division while remaining in the same axis of division. Thus, after having divided animals into footless and footed, one divides the footed into split-footed and not split-footed. Therefore, the "if" of phrase 6 (*ean* with the subjunctive) almost has the sense of "since"; and we should read "if one does not obtain" as if it meant "and it really is the case, as we shall demonstrate," where "one" refers to the person who practices dichotomy.

But I have said that the reference to connection by conjunction is difficult to interpret. Perhaps the two interpretations presented above are useful when taken together. For the recourse to the connection of terms by a conjunction is certainly for the dichotomists a proof of the failure of their method, since that method purports, by just one axis of division, to arrive at a definition. But the result of the last division will not give the sought-for definition: to obtain that, one must recapitulate, and connect by conjunctions, the results of all the preceding divisions. For example, if the last division is between wild and domestic, it does not define the animal; one must add the other divisions and say: tame *and* winged *and* a third characteristic obtained by another division, and so on. By contrast, in a correct division — that is, one that takes the difference of the difference in the same axis of division — the last division will contain the preceding ones and avoid having to repeat them. Thus, the division into split-footed and non – split-footed presupposes the division into footed and footless. To be sure, Aristotle's method will also have to appeal to connection by conjunctions, for the final definition will have to coordinate the results of different axes of divisions: "form of foot," "mode of life," and so on. "That is why it is necessary to divide the unity from the beginning according to several [axes], as we have said" (643b23). But the normal character of Aristotle's multidimensional division is unwelcome in the one-dimensional diairetical method of the Platonists.

Thus arises a problem that occupied Aristotle's attention in *Metaphysics* Zeta 12: if one divides according to several differences in order to construct a definition, why is the unity of the object defined not damaged?

> The aporia is the following: what constitutes the unity of that of which we say that the formula is a definition, as for instance, in the case of man, "two-footed animal"; for let this be the statement

about man. Why, then, is this one, and not many, viz., "animal" *and*
two-footed?   *(1037b10, Ross trans. amended)*

In our passage in the *Parts of Animals,* Aristotle builds upon a
comparison with a sentence that remains one, despite the articulation of its various parts, through the subterfuge of "conjunction." That conjunction has the function of unifying statements
is often affirmed by Aristotle; for example, in the *Rhetoric,* in this
remarkable formula: "Conjunction makes into one that which is
multiple" (3.12.1413b32).[22] But although it is suggested by the
texts, a consideration of our passage of the *Parts of Animals* in the
light of the question as posed in the *Metaphysics* is made difficult
by Aristotle's assertion elsewhere in the same work that the
unity obtained by conjunction is much too loose to characterize
a definition: "A definition is a set of words that is one, not by
conjunction, like the *Iliad,* but because it deals with one object." [23] It must be admitted that our passage of the *Parts of
Animals* is less demanding, for the unity of the thing defined,
than the texts cited from the *Metaphysics.* In the *Parts of Animals,*
once Aristotle has gotten rid of the hypothesis of dichotomy, he
is in fact satisfied with coordinating several characteristics obtained by several axes of division. Indeed, there is no reason for
him to raise here the metaphysical problem of the unity of the
definition, since his objective is to show that his method is
superior to that of the Platonists, in that it can approach a definition of the animal from several directions. But whatever can do
more can also do less. If the true definition, allowed by the rules
promulgated in the *Metaphysics,* has more unity than the definition by conjunctive aggregation of characters, *a fortiori* it has this
much unity.[24]

Thus, beyond the criticism of Platonic dichotomous division,
the passage in the *Parts of Animals* applies a method of division
that Aristotle considers correct. But we may nevertheless be
surprised at the extreme allusiveness of Aristotle's recommenda-
tions concerning the use of his own method of division; only
with extreme difficulty can they be disentangled from the fabric

of this passage. Why does he not give us a formal exposition of his own method? The hypothesis of a lacuna in our text of *Parts of Animals* 1 is not very plausible; there is no room, between the end of the condemnation of dichotomy (chap. 3) and the reminder that the distinctions made by common sense are adequate (beginning of chap. 4), for an exposition of a properly Aristotelian diairetical method. I think there are two reasons for Aristotle's reticence. A fairly simple reason is that it is not necessary to explain something everyone already understands — Aristotle is, on the whole, satisfied with the distinctions made by ordinary people, and those do not need detailed explanation. But the real reason for the absence of a presentation of his *methodology* of division cannot be fully understood until the end of the present study — that is, until after I have presented the status (or place and function) of classifications in Aristotle's biological work.

Finally, the last stage, according to Lloyd, of Aristotle's development of a method for classifying animals is to be found, as far as the biological works are concerned, in the picture he draws of living things in the beginning of the *Generation of Animals* 2, following their modes of reproduction and the perfection of their offspring. I must begin by calling attention to a problem I put to one side during my earlier comments on Lloyd. As it happens, Lloyd first examines the development of the theory of the classification of animals outside the biological works (in the *Organon,* the *Metaphysics,* and the *Politics*), and then in the biological texts themselves. But in the first group of texts, Aristotle's definitive doctrine about classification would be a combinatory method that Lloyd believes is presented in the passage he cites from the *Politics.* As it is totally senseless, no matter what uncertainty exists regarding the chronology of the corpus, to maintain that all the biological texts as a group precede all the other texts, one has the right to ask why there is no trace in the zoological texts themselves of the method of classification developed in the *Politics.*[25] That would be true even if the text cited from the *Politics* were to be late. Research on the chronology of Aristotle's works has not shown that to be false, but, for the sake of coherence, Lloyd's interpretation absolutely demands that late dating, since he sees in this text a last stage of development.

The only remaining solution seems to be a possible boundary between the "logical" and the biological writings *and concepts.* Many interpreters assume a boundary of this kind without admitting they have done so, for it is inadmissible. They establish separate senses for the overdetermined terms of Aristotle's philosophical vocabulary: thus, for the term *eidos* there is a logical sense, a metaphysical sense, a biological sense, a political-ethical sense, and so on. Perhaps that is demanded of the author of an index, and we have to recognize the impossibility for translators of rendering a term like that with just one word. But all the same, interpreters very often allow the unity of the term to dissolve into a more or less accidental homonymy. This procedure for masking textual contradictions is quite comparable to the appeal, attacked above, to an evolution in the author's thought.

But Aristotle himself forbids us to trace a boundary between biological and nonbiological texts, since in the *Politics* he explicitly and at length uses the example of the definition of *animals.* Furthermore, if we play Lloyd's own game — if we look for the stages of Aristotle's possible development in relation to the classification of animals — his position becomes more and more precarious to the extent that presumptions become stronger that the *Generation of Animals* was composed rather late.[26]

Continuing our examination of Lloyd's theory: starting from the undeniable textual fact that Aristotle, in *Generation of Animals* 2, orders animals according to the perfection of their offspring, Lloyd concludes that this new classification abolishes previous classifications, particularly that by organs of locomotion. At first glance, the text might seem to support this reading: at the start, in fact, Aristotle recognizes the impossibility of constructing families that combine in any regular way a mode of reproduction and a mode of locomotion, because, as he says:

Much overlapping of kinds occurs. For the bipeds are not all viviparous (for birds are oviparous) nor all oviparous (for man is viviparous); and the quadrupeds are not all oviparous (for horses, oxen,
and countless others are viviparous) nor all viviparous (for lizards,
crocodiles, and many others are oviparous). Nor does the possession
or nonpossession of feet differentiate them; for there are viviparous
footless animals such as the vipers and the selachians, and oviparous
ones such as the fish and snakes other than vipers.

(732b15 – 23, Balme trans. amended)

It would seem that here, following the logic of the results obtained from our examination of *Parts of Animals* 1.2 – 4, Aristotle
is trying to define classes by a composition of two axes of division (reproduction and locomotion). In fact, these attempts at
correlation will not get us closer to the real definition, because
they cut too close to the perceptual data; but as I shall soon show,
the true definition is causal. Thus, Aristotle looks behind the
appearances for the causes of the differences in the modes of
reproduction, specifically to the degrees of warmth and humidity of the animals and to the combinations of these two principles. From these combinations, Aristotle derives a *hierarchical*
classification, relying on the principle that the perfection of an
animal is proportionate simultaneously to its degree of heat and
its degree of humidity. Consequently, there are five possibilities:

- Vivipara, which give birth to qualitatively perfect young
  (733a33).

- Ovovivipara ("they are viviparous after having been oviparous" [733b5]), which do not give birth directly to perfect
  products, but do have living young.

- Ovipara that lay perfect eggs (733b6).

- Ovipara whose eggs complete their development outside
  the animal, "as is the case for scaly fish, crustaceans, and
  molluscs" (733b9).

- The coldest animals, whose eggs are formed outside; for
  example, insects, whose larvae, Aristotle believed, become
  eggs (733b11).

In fact, if this classification is new, that is because it is adapted to the subject matter of the *Generation of Animals,* which is also new. The specificity of the one comes from the specificity of the other: Aristotle constructs the classification that he needs for his purpose. Lloyd did not understand this because of an unseen and blinding prejudice that makes his reading of Aristotle an heir of the anachronistic interpretations denounced at the beginning of this study. Lloyd implicitly assumes that Aristotle was aiming at *one* definitive classification, and that the several classifications of animals found in his works can only be stages toward this ultimate classification. For Lloyd, too, therefore, Aristotle had a taxonomic project related to that of the systematicians of the modern era.[27]

Until now, I have been presenting the difficulties brought about by previous interpretations of the relationship in Aristotle's writings between the division and the classification of animals. A literal reading of the texts is enough to refute these interpretations. The most rapid examination shows, in fact, that Aristotle did not condemn divisions in zoology, and that his criticism of the dichotomic method was actually aimed at establishing his own diairetical method more firmly. Thus, there is not, in the biological corpus as we know it, any decisive alteration in Aristotle's attitude toward division, despite the probable temporal spread of the biological works. If, then, Aristotle remained a convinced "diairetician," we cannot in any way attribute to him an eventual failure of his taxonomic enterprise brought about by the liquidation of the theoretical tool for classification that is *diairesis.* Ultimately, we are dealing with a problem produced out of nothing by the commentators themselves.

When we look at Aristotle's critique of Plato, we must first get rid of the immediate and false idea that the attack on division in zoology was only a local skirmish in the more general epistemological war on division waged in the logical and metaphysical works. In other words, along with reevaluating the role of division in Aristotle's doctrine of knowledge, we must restore diairetics to the place that Aristotle always saved for it in zoology. In fact, as I have shown elsewhere (Pellegrin 1981), for Aristotle to say that division does not demonstrate — that it is, as Aristotle says, "like an impotent syllogism" (*APr.* 1.31.46a33) and that ultimately it is not an argument, in that it yields no conclusion — is in no way to eliminate it from the theory of knowledge and of scientific practice. One could even say, on the contrary, that in constructing definitions, division becomes a possibility-condition of scientific reasoning (syllogistic), since Aristotle counted definition among the principles. Resisting once again the attempt to establish a boundary between logical and biological senses of words, we will see that in zoology division keeps its definitional function, and that it was in respect to the definitional capacity of division that Aristotle's criticisms of Plato were elaborated. Thus, the quarrel between them was not that of rival taxonomists: Aristotle did not blame his teacher for having failed to develop a rigorous classification, and consequently he did not try to give a lesson in taxonomy to the Platonists.

To see in the criticisms that Aristotle aimed at Platonic dichotomy a debate between taxonomists is to fall into the anachronism that we outlined above. Neither Aristotle's reference to the "classificatory tables" used in the Academy (*PA* 1.2.642b12; *GC* 2.3.330b16) nor the witness of Athenaeus should make us forget Plato's goal in the dichotomous method of division. To pretend that Aristotle accused him of being a second-rate classifier is to imagine that he attributed to his teacher the goal of classifying. Aristotle could not (yet) fall into that anachronism.

Plato undeniably posed problems that have proven to be stumbling blocks for all who have tried since the Renaissance to establish "natural" classifications of living things. He did indeed show that the *point of view* from which distinctions are made places a preliminary obstacle before the diairetician. We need only recall on this question the passage in the *Statesman* (263c) in which the Stranger mocks the haste of the Young Socrates to divide animals into human and non-human, pointing out that if we were to ask the cranes, they would divide animals into cranes and non-cranes. But the persistence of certain formal problems across systems of thought that are so radically different as Plato's and that of the classical taxonomists is not sufficient to show that Plato was a systematician.

Unquestionably, Plato's goal was not classifications but definitions. In the *Sophist* and the *Statesman,* he assigns two tasks to division: (a) to allow participants in the dialogue to move from agreement about the name to agreement about the thing; and (b) to unmask people (e.g., sophists) who pretend to be something they are not.[28] We must, however, notice that although Plato indeed aims at definition by means of division, he succeeds only in the case of the angler. In the case of the sophist, several diairetic procedures, none rejected by anyone in the dialogue, fail to give an acceptable definition: a long detour by way of being, non-being, and the "great kinds" is required before the participants in the dialogue arrive at the definition sought. As for the statesman, the division by itself fails to distinguish between the legitimate and the illegitimate claimants: for that, a detour by way of the theory of the right measure and recourse to the myth and the paradigm of weaving are needed. But however that may be, far from aiming at the establishment of a taxonomy with an intrinsic theoretical value, Platonic dichotomic division is only a tool, or perhaps we should say a weapon, in Plato's metaphysical strategy in the critique of knowledge based on appearances.

Far from attacking his teacher for not having constructed a *systema naturae,* which would certainly be a systematist's criticism, Aristotle bases his case on the definitional capacities of division. The sentence that begins his criticism of Platonic division reads: "Some [members of the Academy] obtain the particular by dividing the genus into two differentiae" (*PA* 1.2.642b5, trans. Balme). Aristotle has thus observed that Plato's dichotomic divisions aim at isolating the particular, and that would be a strange procedure for a taxonomist. However, he does not attack the *purpose,* but the *method,* which he finds "simultaneously not very easy and impossible." [29] Thus, when Aristotle disapproves of dichotomy's separating natural groups, we are tempted to believe that he is accusing it of an inability to constitute "natural" families of animals, when in fact he is taking it to task for impeding the correct understanding of individuals: for example, putting swamp birds with fish among aquatic animals obscures their real nature and leads us away from their definition. We shall see, in a later chapter, that Aristotelian biology ultimately does not aim at definitions of what we call species of animals. Nonetheless, animal species have a real nature that the zoologist should make manifest. But that is true of any group of animals, no matter what the degree of generality on the taxonomic scale. "Bird," for example, has an *ousia* (see *PA* 3.6.669b11; 4.12.693b13). But species of animals indubitably have an "immediate" obviousness, and thus they must be found at the beginning of the process of biological research. Thus, in *History of Animals* 1, after a list of many points of view according to which animals could conveniently be distinguished, Aristotle writes:

> This is why we must consider the nature of each of them separately. Proceeding in this way can, for the moment, be called an outline, a foretaste of the way in which one ought to carry on the investigation and about what the investigation should be carried on.
>
> *(1.6.491a4)*

This necessary consideration of the nature of animal species sometimes gives first-time readers of the *History of Animals* the impression that they are dealing with a catalogue of brute facts in no particular order; that impression is false, as I shall show.

Not to anticipate too many of the future results of our inquiry, let us be satisfied for the moment with the reaffirmation that Aristotle aims his anti-Platonic polemic at the definitional value of the diairetical method. This point should not be doubted by anyone in either camp, Plato's or Aristotle's.

Permit me a more general remark on this topic. Between
Plato, Aristotle, and contemporary commentators, there are
major caesuras whose locations vary according to the vantage
point one adopts. In the first place, there is a gap between Plato
and Aristotle that we are doubtless in a better position even than
Aristotle to appreciate. As has often been demonstrated,[30] Aristotle is not an historian of philosophy in the modern sense;
besides, it is a general rule that temporal and ideological proxim-
ity obscures veridical perception of differences.[31] The Aristotle
who, in his criticism of Plato, takes Aristotelianism as the only
possible theoretical framework is the worst possible historian of
Platonism. Perhaps it is too much to demand of an innovator
that he step outside himself and pour himself into the mold of
another person's thought. But there is another gap, one that
separates Plato and Aristotle together from modern interpreters;
and as a result of this, Aristotle is the best possible historian of
Platonism, in that the two philosophers belonged to the same
intellectual world, which is radically foreign to us. What Aristotle criticizes in Plato is ultimately not so important for us,
since those criticisms most often miss Plato's point and so are
often more useful to the historian of Aristotelianism than to the
historian of Platonism; more important is what he admits without discussion. In the group of propositions that are "not a
problem" for the people of a period, one discerns most surely the
ideological foundations of that period.

It is therefore very important for us that Aristotle locates his
anti-dichotomic polemic with the Academy not on the level of
taxonomy (which is the reading of modern interpreters) but on
that of definition. Ultimately, it does not matter very much for
us whether Aristotle's criticisms are or are not based on a correct
understanding of Platonism; what we are sure about is that the
stakes in this contest are "metaphysical" and have to do with the
ability of division to define.

If, then, Plato and Aristotle are in agreement in assigning to
diairesis a definitional function, what basic criticism can Aris-

totle aim at dichotomy, from the point of view of its definitional value?

Their disagreement starts from the status of definition itself. In fact, Aristotle would never consider a reality as defined simply on the ground that it has been distinguished from everything else, even though such a distinction may have removed all possible ambiguity concerning the thing defined. But Platonic diairetics, as we have seen, is a process of identification by successive distinctions meant to prevent something from appearing to be what it is not. Aristotelian definition, in contrast, claims to get to the ontological structure of the thing defined. That is demonstrated both by the Aristotelian thesis that every good definition is causal and by the frequently repeated assertion that there are no real definitions except of *ousia.* We must review these two theses to the extent that they influence Aristotle's biological point of view.

Commenting on the statement "Definition is a formula showing why the thing is," found in the *Posterior Analytics* (2.10.93b39), Gilles-Gaston Granger properly writes (1976, 236):

> Definition is "a formula making manifest the 'why' of the thing."
> While "definition" in the first sense only designates, signifies, in
> this second sense it presents the cause *(dia ti)* and is then "like a
> demonstration of the essence." Not that the definition could by
> itself demonstrate the existence of the thing: one should rather say
> that it lays bare the existence-condition. Apparently, that is what
> Aristotle shows with his example of thunder.

Definitory discourse, whether concerning animal kinds or something else (we shall return to this problem later), because of its quasi-apodictic nature, cannot be a simple process of isolating (a sparrow is not a crane), but must be a real account, that is to say, an articulation of parts establishing the necessity of the nature of the thing considered. "To define," writes LeBlond (1939b, 353), "is not to apprehend the essence, it is to explicate it."

On the other hand, and for him this is in no way in opposition to the preceding aspect, Aristotle constantly repeats that the definition is a formula that expresses the essence (*ousia* or *to ti en einai*).[32] It would hardly be possible to take up here the whole question of the relationship between definition and essence; I shall touch on it only in connection with something I observed above in relation to the critique of dichotomy, namely, that the true (i.e., Aristotelian) diairetical method ought not to divide according to accidents. That there is definition only of *ousia,* and not of accidents, that division permits the construction of an adequate definition, and that a good division does not divide according to accidents, all converge upon the same doctrine. But the Aristotelian theory of accidents operates on two levels of unequal rigor. On the first level, Aristotle opposes the accidental to the "per se." [33] An accidental attribute would be something that cannot be said of a subject either necessarily or even usually, so that, according to the famous example of the *Metaphysics* (Delta 30.1025a15), to find a treasure when planting a tree is an accident. Similarly, it is not *qua* biped, but accidentally, that an animal is tame or wild. But Aristotle proceeds to another level when he speaks of "per se accidents." [34] This expression has given some trouble to commentators, and their embarrassment is reflected in their translations. Thus, Octave Hamelin translates *symbebekos* sometimes by "accident" and sometimes by "attribute." [35] The passage that follows the text of the *Metaphysics* mentioned above will show sufficiently what Aristotle meant by "per se accidents":

> "Accident" has also another meaning, i.e., all that attaches to each thing in virtue of itself but is not in its essence *(ousia),* as having its angles equal to two right angles attaches to the triangle. And accidents of this sort may be eternal, but no accident of the other sort is.
>
> *(Delta 30.1025a30, Ross trans.)*

What interests us here, as we investigate division, is that the doctrine of per se accidents requires greater rigor from diaireticians in their practice: it is insufficient, when one constructs the definition, to distinguish only the accidental from the per se, since the per se must itself be accidental when it does not belong to the *ousia,* and consequently falls outside the definition.

One might well ask about the properly biological use and import of this very rigorous theory of per se accidents. The doctrine reappears, with this name, in *Parts of Animals* 1, but it seems at first sight to be applied ambiguously. Aristotle uses the word *symbebekos* here in two different senses, apart from its meaning of "accident" as opposed to "per se." [36] First, we find a passage that agrees so completely with the one from the *Metaphysics* quoted above that it gives the same example:

> Further, one should divide *(diairein)* by what is in the being *(ousia),* and not by the essential accidents — as if one were to divide figures on the basis that some have their angles adding up to two right angles and others to more; for it is an accident of the triangle that it has its angles adding up to two right angles.
>
> *(3.643a27, trans. Balme)*

Here, then, there is no gap with the metaphysical doctrine of definition. In zoology, too, correct division should lead to a rigorous definition; and to do that, it should limit itself to predicates included in the essence.

But another passage in *Parts of Animals* 1 seems to recommend something entirely different:

> It is necessary first to divide off, in relation to each kind, the attributes that belong essentially to all the animals, and then to try to divide off their causes. *(5.645b1, trans. Balme)* [37]

Here, too, what is at issue is division, as the aorist *dielein* indicates, and not description, as Peck says. The per se accidents in question, as the immediately subsequent text makes quite clear, are the "parts" *(moria)* that, in related or analogous forms, belong to all animals; Aristotle gives as examples "feet, wings, scales" (645b5). [38] But any contradiction with the previous passage disappears if we notice that Aristotle is not inviting us to divide *according to* the accidents, but to divide the per se accidents themselves. What this passage says in addition to the previous one is that these per se accidents are objects of investigation for the biologist. It is Aristotle's permanent position that the per se accidents, even though outside the essence, and thus not parts of the definition, are nonetheless objects of scientific research. In a passage in the *Posterior Analytics* which establishes that there is only one science per genus, Aristotle writes:

> For there are three elements in demonstration: (1) what is proved, the conclusion — an attribute inhering essentially in a genus; (2) the axioms, i.e., axioms which are premises of demonstration; (3) the subject-genus, which attributes, i.e., essential properties, are revealed by the demonstration.   *(1.7.75a39, trans. Mure)*[39]

In the passage of the *Parts of Animals* that I have been discussing, the project is to divide the genus "foot" or "wing" or "scale" into different species that differ according to the animals. I must insist on this point, because most interpreters and translators have gone astray: there is in this passage a diairetical recommendation, and Aristotle does not say that it should be done according to accidents. We shall see farther on that this recommendation, and all those that resemble it, are applied punctiliously when — for example, in the other books of the *Parts of Animals* or in specialized treatises like the *Progression of Animals* — Aristotle studies the various forms that the organs of locomotion can take, or the organs of digestion, and so on, in various animal species. After this, he proceeds to distinguish the causes of these various organic forms.

This interpretation is again confirmed by an earlier passage of the *Parts of Animals* 1, delimiting the biologist's field of study:

> If, then, this is so, it will be for the natural philosopher to speak and know about soul (if not about all soul, then about just this part in virtue of which the animal is *such*); he will both say what the soul is (or just this part of it) and speak about the attributes that belong to the animal in virtue of its soul's being such [literally, the text says "the accidents which are those of the *ousia* of this soul"].
>
> *(1.641a21, trans. Balme)*

It is the soul that is the essence of the living thing, or at least a part of the soul: we know the hesitations and doubtless changes in Aristotle's doctrine on this point. And the natural philosopher should occupy himself with the accidents of this essence: obviously, Aristotle is talking about per se accidents.[40]

The critique of dichotomy leads us, therefore, to another use of division. But this use of division does not result in the construction of a taxonomy: by the intermediary of definition, division leads us to the cardinal Aristotelian metaphysical concept, that of *ousia*. Not that the zoologist has to ask the question, "what is substance?" for that is the business of the "first philosopher." But zoology is, in its diairetic part, "ousiological." It constructs the definition — that is, it arrives at the "formula of the essence"—by dividing the essential attributes. Division must therefore go beyond the immediately given attributes, even if they are per se, to discern, through the true differences, the nature of the animals, and especially, as we shall see, the nature of their "parts." [41] But dichotomy, which divides according to any characteristic at all, misses the level of essence. For example, it would put man and bird together in the class of biped. But that is, in both cases, a per se accident (for all men are bipeds), and as such, this per se accident can be put in opposition to "those accidents that are not per se . . . and of which there is no demonstrative science" (*APo.* 1.6.75a18) — as, for example, in the case of a man, the fact of being seated. But, as the *Parts of Animals* notes, in man and bird "bipedicity is other and specifically different" (1.3.643a3).

At first glance, modern taxonomists, too, seem to condemn immediateness, in that they try to construct an *objective* classification of animals, a classification free of the subjectivity of the individual taxonomist. But as a matter of historical fact, it was only by the elimination of the ousiological goal that taxonomy came into being. Its development depended upon a renunciation of the desire to grasp the intimate nature of the living thing. Instead, taxonomy reduces the living thing to a collection of characteristics.

This recognition of the fundamental role played by the question of *ousia* permits us to reexamine, on a firmer basis, a fact already noticed but often misunderstood by interpreters: Aristotle's preference for "common sense" rather than the Platonic dichotomic method. The reasons alleged for this by some commentators are not necessarily wrong. For example, Louis Bourgey (1955, 126) makes at least two good points. In the first place, "to take account of already-recognized groups [of animals] that the long experience of professionals (fishermen, hunters, stockmen) has constituted seems infinitely more certain" than the Platonic method; and, Bourgey adds in a note, "it is important to emphasize that Aristotle starts from the extensive experience of an observant seafaring people." Certainly, the paradoxical subtlety of dichotomy would not have facilitated contacts between the theoreticians of the Academy and practical specialists, thus depriving the Academics of an important source of information. In contrast, one finds in Aristotle traces of information taken directly from specialists, which Georges Pouchet arbitrarily interprets as a proof of inauthenticity: for example, he thinks that only a beekeeper could have written the chapter on bees in the *History of Animals* (9.40).[42] It is much more likely that Aristotle had observed the beekeepers of whom he speaks several times and whose operations he describes fairly closely.[43] We should note that Aristotle's undeniable concern for direct observation ultimately hurts more than it helps him today; while we like to recognize in him a certain affinity with modern experimental procedure, his serious errors, which *we* think could have been avoided by simple observation, continually cause great perplexity for our contemporaries. We should also recognize that Bourgey is justified in saying (1955, 125):

> The naturalist's starting point is not some ingenious mental construction, but is much more likely to be the framework furnished by ordinary thought; but Aristotle does not accept these frameworks indiscriminately just as they are; he works on them to distinguish between them, to rectify and to refine them.

I have already mentioned such distinctions, following Michael of Ephesus. It is even more interesting to notice that Aristotle himself supplies an epistemological warrant for what he is doing, as we can see from the distinction he makes between the respective viewpoints of the natural philosopher and the practical man. Thus, in *Generation of Animals* 3.5.756a33, he writes that the sexual congress of fish could have escaped the notice of fishermen, otherwise well-informed about the habits of aquatic animals,[44] because of the speed with which it occurs, but especially because one finds in fishermen no theoretical objective that could have pushed them to a precise observation, "for none of them would observe that for the sake of knowing."[45] Again, with a precaution that recalls Herodotus,[46] Aristotle often notes that he has not verified a fact himself, or at least he has not had it verified by a trustworthy person, that is, by someone with the same "historical" viewpoint as himself. (We should never forget that both the Academy and the Lyceum worked collectively.) Thus, he writes about the sponge in the *History of Animals:*

> It seems that the sponge has some sensation, for it is said that if not approached with care, it is more difficult to pull up. *(1.1.487b9)*[47]

But there is a deeper reason for this critical resort to what "is said": that is, that ordinary thought is spontaneously "ousiological," and that in this sense it approaches truth more closely than the dichotomous method can. Thus, among the many texts that one could call upon here, we read in the *Metaphysics:* "It seems that *ousia* belongs most obviously to bodies; and so we say that animals and plants and their parts are *ousiai*" (Zeta 2.1028b8). "It seems" and "we say" obviously designate current opinion. What Gaston Bachelard calls the "substantialist obstacle," which sciences must surpass in order to constitute themselves or to develop, is for Aristotle a viewpoint that is "scientific" ("epistemic" in the etymological sense). The spontaneous zoology of the man in the street (and the first example that Aristotle gives of substances recognized as such by everybody is precisely that of animals and their parts) is thus less distant from a truly scientific approach (i.e., the ousiological problematic) than is the Platonists' perspective.

We have thus arrived at the following results: Although

Aristotle condemns dichotomy as used in the Academy, and does so in all the relevant texts, he does not reject division.[48] However, he does not give division the task of constructing a taxonomy of animals; not only does he not take Plato to task for having failed to develop such a taxonomy, but, as a matter of fact, he never raises a classificatory question in a way that would call upon the use of division. On the contrary, in his zoological as in his "logical" texts, division leads to the essence by way of the definition, which is the "formula of the essence."

The question that now arises is: how does Aristotle divide? And more particularly, how does he divide in order to arrive at the zoological definitions he is searching for? Our examination of this question will make the problem of taxonomy reassert itself, because the Aristotelian procedure utilizes the concepts of *genos, eidos,* and *diafora,* which translators generally render by "genus," "species," and "specific difference." Let us see, then, briefly, before studying them for their own sake, what relationships these concepts have to definition, the goal of division.

To do that, we can reexamine the results presented in an article, already cited, by LeBlond (1939b), who notes that one may approach definition from three directions (unfortunately, he eventually writes that these different "perspectives" constitute "three kinds of definition"[49]—which certainly goes beyond his original idea). To define is, in the first place, to distinguish the matter and form of a substance and to grasp their unity; this procedure "suits substance and constitutes definition *par excellence*" (p. 364). In the second place, it is to "grasp the duality of *cause and effect*" (p. 358). "Let us finally note a third schema of analysis-synthesis that definition allows: definition according to *genus and species*" (p. 358). These three relations have in common that they are all relations of potentiality to actuality (p. 359). LeBlond's study follows three stages. After having tried to clarify the *nature* of definition, he examines the *method* that allows it to be achieved (here, too, he retains three procedures: a kind of induction, Platonic division, and the "essential syllogism" as presented in the *Posterior Analytics* [2.8]); and finally, he examines the *place* of definition.

But the truth is that LeBlond does not seem to respect his own plan. He should have put definition according to genus and species among the methods, instead of "Platonic division," which — he must have realized — Aristotle had rejected. In fact, we cannot help seeing in this curious appeal to Platonic method an obscure and badly formulated recognition of what I established above — namely, that Aristotle retains division and gives it a role in the construction of definitions. Division operates by means of "genus" and "species" in constructing definitions; this passage, among others, from the *Metaphysics* demonstrates that:

> For definition is a certain formula that is one and that is about substance and essence, so that it must be a formula about some one thing, for substance signifies something one and individual, as we have said. And, in the first place, we must inquire about definitions reached by the method of divisions. For there is nothing in the definition except the genus called "first" and the specific differences. *(Zeta 12.1037b25)*

This passage reminds us, first of all, that a definition is the formula of the essence, and that division, which is here the properly Aristotelian method of division, and not Platonic division as LeBlond seems to believe, is a means of constructing a definition. I have shown elsewhere (Pellegrin 1981, 181) that Aristotle even recognizes, if only by omission, that division is the *only* means of constructing a definition: the expression "in the first place *(proton)*" at 1037b28 is not, in fact, followed by any "secondly." But this passage also asserts that in order to arrive at a definition, division works on the "genus," which it divides according to its "specific differences." This division of the "genus" into "species" is thus really a *method* of obtaining a definition; or rather, the Aristotelian diairetical method well and truly consists of dividing the "genus" into "species." Therefore, we must now examine the three notions of *genos, eidos,* and *diafora.*

2

GENUS, SPECIES, AND SPECIFIC DIFFERENCE FROM LOGIC TO ZOOLOGY

FOR THE BEST POSSIBLE UNDER-standing of the internal coherence of Aristotle's investigations, we must try to rely on a methodological postulate that may be stated thus: there is no strict line of demarcation between Aristotle the metaphysician and Aristotle the biologist. This postulate of the conceptual unity of a corpus in its various aspects ought to be the basis of any study in the history of philosophy; but it is especially difficult to remain faithful to it in the case of Aristotle, since his work seems so fragmented. Furthermore, relying on this postulate does not mean forgetting the relative theoretical autonomy that biology acquires from the character of its object: because biology concerns itself particularly with living things, terms applied to living things have a specifically biological use. But we must keep in mind that even if there is a properly biological *use* of the words *genos, eidos,* and *diafora,* we cannot talk about a biological *sense* of these words. That does not mean that we must consider the use of these concepts in the biological corpus as a simple *application* in the domain of living beings of a conceptual structure developed in the "logical" texts. We cannot do so partly because the different "branches" (logic, metaphysics, biology, etc.) of Aristotle's work are not successive stages of his development. But we must also be concerned here with a general problem in the interpretation of Aristotle, that of the theoretical affiliation between logic and biology. Although I have not yet presented all the means for resolving this problem, I must say something about it, since so many commentators, including some of the best, have proposed solutions that are too hasty and thus suspect.

Jean-Marie LeBlond (1939a, 72), for example, considers the problem of the relationship between logic and biology resolved, without need for much discussion, because syllogistic reasoning has "a biological origin." He seems to have been convinced by the following argument: (a) "The syllogism," he writes, "is thus definitively explicated by the relation of included to includer that exists between its terms." (b) Aristotle envisages relations between the terms of syllogisms from two points of view, which LeBlond calls "comprehension" and "extension," explaining that those terms are approximate since anachronistic. (c) The extensional point of view leads Aristotle to compare *classes,* and "it is principally according to this second formulation that Aristotle establishes the rules and the figures of the syllogism." (d) But Aristotelian biology is based on a practice and a theory of the fitting together of classes of living beings; if LeBlond does not say so explicitly here, that is simply because for him it is obvious. He therefore concludes:

> The discipline that serves to guide Aristotle's logic, that inspires at least his explanation of the syllogism, is, as strange as it might at first appear, biology.[1]

However, the idea that in biology Aristotle applies schemas developed in the metaphysical and logical works implies a solution for our problem that is just the opposite of LeBlond's. Thus, Abel Rey writes:

> The classification that is at the basis of natural history is nothing but the application of the logical scheme of the *Prior Analytics,* the achievement of the whole evolution of Greek thought toward the rational since the polemicists and the first mathematical demonstrations at the end of the sixth century.[2]

We can as yet only raise the question of the relation between Aristotle's logic and his biology. But it is better to ask it while realizing that we cannot for the moment answer it than to decide the question without having realized that it has been asked. In any case, it will arise again, since it is intimately related to the question of taxonomy. In fact, to try to establish a biological ancestry for Aristotle's logic, LeBlond relies on the hypothesis that Aristotle's zoology is based on an ordering of living things into at least approximately stable and constant families.

My postulate is less radical but in fact more difficult to use: I do not believe we can find any clear order of descent between the logical and biological works, but rather a "simple" theoretical unity that is reflected in the use of identical terms in domains that *we* judge to be heterogeneous. My assertion and defense of the theoretical unity of Aristotle's work could not be astonishing to him, no matter how much theoretical unity is an essentially modern requirement — one that perhaps shows the traces of the autonomous development of the sciences which was fatal to that unity of knowledge that the classical philosophers called philosophy. Once again, it is *we,* or our immediate predecessors such as the systematicians of the eighteenth century, who give a theoretical autonomy to Aristotle's biology by separating it from its metaphysical foundations.

If, then, I begin by examining the logical functions of the concepts of *genos, eidos,* and *diafora,* it is not because I recognize a precedence of logic over biology, but by way of a detour. The use of these three concepts in the biological treatises seems, at first sight, to be characterized by a great confusion. We shall see how the apparently irregular utilization of these terms in the biological treatises makes many commentators lose hope. We shall also see that their attempts to make sense of that irregularity have had results sometimes discouraging, sometimes ridiculous, despite their ingenuity. I hope to reduce the irregularity by circumventing it. We learn the rules of their biological employment by looking at the functions of these concepts in the metaphysical and logical texts, where they are studied for their own sake, independently of any specific field to which they might be applied. And there, where our predecessors have noticed only inattention or approximation, we shall see the application of strict rules of usage.

How, then, can dividing produce an adequate definition? To succeed in his project, Aristotle was led to develop a theory of classes and their interconnections, by articulating the three concepts of *genos, eidos,* and *diafora.* And it is by understanding these terms as "genus," "species," and "differentia" that systematicians of the modern period have made of Aristotle the inventor of taxonomy.

One could hardly deny that there is to be found in Aristotle's texts a precise — one might say "technical"— delimitation of the terms *genos, eidos,* and *diafora.* I shall leave to those better informed, or braver, than myself the task of discussing whether or not Speusippos preceded Aristotle on this path.[3] The commentators, including the best among them on this point, such as David Balme (1962a, 83), generally conclude that this technical sense and a nontechnical sense coexist in Aristotle's texts. They sometimes rely on this distinction to rid themselves of a bothersome text by deciding that it uses a nontechnical sense. I have already called attention to the problem of the coexistence, for one term, of several "senses." It is particularly dangerous in the case of the concepts we are now examining. That appears clearly in the divisions of the Bonitz *Index:* before considering the word *genos* "in its logical sense" *(logice),* the first paragraph gathers the uses with the idea of *proles, familia, gens* with some derived senses, as those of sex (doubtless a bad reading, as we shall see). Similarly for *eidos,* which can mean: (1) *externa figura ac species;* (2) species *logice;* (3) the Platonic "ideas"; (4) the "formal cause."

Let us rid ourselves, once and for all, of chronological explanations of this semantic diversity. Certainly, there is an initial temptation to resort to such an explanation, since we may take it for granted that the terms *genos* and *eidos* were relatively interchangeable for Plato.[4] That granted, should not the terminological flux in Aristotle belong to the early "Platonizing" works, where the two concepts should be used in the vague sense of "kind," while the "mature" works would make them precise by putting them into the hierarchy of "genus" and "species of the genus"? The texts refute such an interpretation. Let me give two examples: In the *Meteorologica,* at 3.6.378a20 and 22, Aristotle speaks of two "exhalations," vaporous and smoky, which two kinds *(eide)* of bodies form in the depths of the earth; the first is formed from metals, the second from minerals. Among the latter, he cites several varieties *(gene)* of stones: realgar, ochre, and so on. We thus see that here *eidos* has a wider extension than *genos.* But from its first page, the *Meteorologica* represents itself as posterior not only to works considered as early, such as the *De Caelo,* but to certainly later writings, such as the *Physics,* the *Generation and Corruption,* and some books of the *Metaphysics.*[5] By contrast, Book 4 of the *Topics,* a work that critics generally recognize as early, neatly delimits the functions of *genos* and *eidos* understood as logical classes. Thus, for their extension:

> In all such cases, the basic principle is that the genus has a wider extension than the species and its difference; for the difference also has a narrower extension than the genus.
>
> *(4.1.121b11, trans. Forster, Loeb edition, with minor modification)*

We must therefore give up trying to discover the moment at which the "technical" use of these terms first arises in Aristotle's work. The problem of the coexistence of senses remains.

But when he asks himself how to divide, Aristotle does not directly respond by construing the two concepts of *genos* and *eidos,* but rather that of "specific difference," which simultaneously presupposes and makes precise the first two.[6] To understand what specific difference is, one must first see what "differ" means. This latter point is well clarified in a passage of the *Metaphysics,* in Book Iota, to which we shall pay special attention, since it is there that the concepts that interest us are characterized in the greatest detail:

> For everything that differs, differs either in *genos* or in *eidos;* in *genos* if the things do not have their matter in common and are not generated out of each other, as are the beings that have forms different in category; and in *eidos* if they have the same *genos.*
>
> *(Iota 3.1054b27)*[7]

The distance between *genos* and *genos* is not of the same nature as the distance between *eidos* and *eidos.* That is made clear a few lines farther on:

> For things that differ in *genos* have no path to one another, but are too distant and without common measure. *(Iota 4.1055a6)*[8]

The intergeneric gap is thus directly impassable: from one *genos* to another, there is no path *(hodos);* i.e., as the text quoted above says, there is not between them the amorphous continuity of a common matter that would permit one *genos* to precede another by generation *(genesis eis allela).*

The term *asymbletos,* which I translate "without common measure," is remarkable. A rare word derived from the verb *symballein,* literally "to throw together," it marks the impossibility of putting two things on the same basis in order to compare them. Tricot translates here by *"incombinable,"* which is doubtless too restricted, and he claims, without giving any reason, that this term must be of Platonic origin.[9] We shall better appreciate its import by comparing this use with two others in Aristotle. In the *Physics* (4.9.217a10), Aristotle says that in a void the notion of movement loses all significance, so much so that "speeds are no longer commensurable." In the *Parts of Animals* (4.2.677a9), in the course of refuting Anaxagoras's idea that in acute diseases the bile invades the viscera, Aristotle answers that that is impossible because these two quantities of bile are too dissimilar—"without common measure," as one may say in English. A. L. Peck translates: "Besides, there is no comparison between the amount of bile which is present in these ailments and that which is emitted from the gall-bladder." In these two examples, then, *asymbletos* marks a relative, even paradoxical, incommensurability, in that it is applied to terms (speeds, quantities of bile) that are ordinarily commensurable. In any case, that interpretation is confirmed by our passage from the *Metaphysics,* which says, "too distant and without common measure," since the word *pleon,* "too distant," although having many nuances according to the context in which it is used, always designates a greater *quantity* of some kind. This presence of quantity, and thus of a possible measure, in the very heart of incommensurability, seems to deny here, at the level of *terms,* what the *content* of Aristotle's statement affirms. Doubtless, this is an indication, whether intended by Aristotle or not, that ultimately the gap between genera is not absolute, and that it will be possible to go from one *genos* to another by blazing another trail. That is why I called this gap *directly* impassable.

While we wait to find that other trail, which will be the path of analogy, it seems that we could give as a fundamental definition (Aristotle would say *haplos*) of *genos* that enclosure beyond which there is nothing but pure otherness, but within which, in its very unity, the diversity of *eide* is produced. In defining *genos,* a unity both identical and diverse, Aristotle often makes use of formulas that ring with a Hegelian sound in our modern ears. Note, for example, these three definitions, again drawn from the *Metaphysics* Iota: "*Genos* means that by which two different things are said to be essentially the same" (Iota 3.1054b30);[10] "For by *genos* I mean that one identical thing that is said of both and is differentiated in no merely accidental way [i.e., that has in it a difference that is not accidental] ";[11] "For I give the name of 'difference in the *genos*' to an otherness that makes it itself other" (Iota 8.1057b37 and 1058a7).[12]

This closure of the *genos* on itself, establishing its unity by deploying its differences within itself, leads also to the doctrine (cardinal in the Aristotelian epistemology) called "incommunicability of genera": that there is no science except within a *genos* (allowing for cases of the subordination of the one science to another).[13] When, in zoology, *genos* designates a group of animals, we shall find again this demand for closure: in the same *genos* will be those animals whose ancestry permits us to say that they are in some sense the same, in contrast to all others. This remark provides a basis for a *theoretical* approach to a part of Aristotelian zoology often studied merely descriptively: teratology.[14] The fact that animal species are deployed in the sublunar world — that is to say, in a cosmic space marked by imperfection and irregularity — will translate into certain hitches to this incommunicability-in-principle; thus, there are some groups that are in a way perverse.[15]

A generic enclosure immediately defines, or rather is defined as, the space of "specific differences": δέδεικται γὰρ ὅτι πρὸς τὰ ἔξω τοῦ γένους οὐκ ἔστι διαφορά (*Metaph.* Iota 4.1055α26). We could understand the expression τὰ ἔξω τοῦ γένους in two ways: first, it can designate "that which is outside the *genos*," and one should then translate the phrase quoted: "for it has been shown that for the things situated outside the *genos* there is no difference," that is, no difference in the *genos* considered. One therefore cannot attribute to an extra-generic being a difference that operates within the *genos* which that being is outside of. Thus, to refer to one example among others proposed by Aristotle: in the *Posterior Analytics* (1.10.76b7), arithmetic assumes differences such as odd and even, which cannot be referred to geometrical figures. But one can also understand τὰ ἔξω τοῦ γένους as signifying, under a dense form habitual to Greek and particularly to Aristotle, "the things situated on both sides of the boundary of the *genos.*" The phrase of the *Metaphysics* then means: "for it has been shown that for the things situated on both sides of the generic enclosure, there is not between them a difference." In other words, things cannot differ (in the sense of "having a specific difference") except within the limits of their generic identity, so that a number cannot differ from a figure from the point of view of evenness.

Diafora is therefore a *determinate* blend of the same and the other: it posits the other because it has the function of differentiating, but it cannot exist except when sheltered from otherness in the heart of the identity of the *genos*. That is what Aristotle has indicated without possible ambiguity in distinguishing specific difference from otherness:

> But difference and otherness are distinct. In fact, on the one hand, that which is other and that than which it is other do not necessarily have to be other in virtue of something [they share]; for all that is, by the very fact that it is, is either the same or other; that which is different, however, is different from something by something, so that it is necessary that there be a same by relation to which the different differs.[16]

As this passage continues, it shows that this "same," by relation to which the different differs, "is the *genos* or the *eidos*, for all that differs, differs either according to the *genos* or according to the *eidos*" (1054b27): thus, difference vanishes beyond the *genos*, where there is nothing left but pure otherness. Within the *genos*, difference is made with respect to *eidos*: "There is no difference with respect to *eidos* except of things in the same *genos*" (Iota 3.1054b30). It is in this sense that we must understand the thesis that even Aristotle's "early" texts supported (e.g., *Cat.* 13.15a4; *Top.* 4.2.123a14), whereby the *genos* is prior to the *eidos*: the first is the possibility-condition for the second.

Thus, the canonical formulation of the relations between the three terms *genos, eidos,* and *diafora:* "The *eide* issue from the *genos* and from its differences" (*Metaph.* Iota 7.1057b7). Here, too, we are dealing with an "early" doctrine in Aristotle.[17]

Both pre-systematicians and systematicians have thought that they could derive from this triple relationship an Aristotelian ancestry for their taxonomic enterprises. In fact, they believed that they had taken their conceptual tools from Aristotle's hands. An assertion that a taxonomic enterprise could have developed out of Aristotle's concepts would not be utterly fantastic, since that is precisely what did happen between the sixteenth and eighteenth centuries. Between Cesalpinus, Tournefort, and Linnaeus there is on this point a true continuity, for they explicitly rely on each other, and the first of them on Aristotle; all of them made of *genos* and *eidos* purely taxonomic terms. The classificatory procedure was constructed, first in botany, step by step in the following manner: To begin with, it was necessary to fix the *genera* of plants; we know that for Linnaeus the "organs for fruiting," in their various arrangements, defined the different genera. Then the *species* were determined by the variations in each of these arrangements. Finally, the systematicians went beyond the borders of the pair genus/species both upstream and down to find *classes* and *orders* in one direction and *varieties* in the other. In zoology, things were to be a little more complicated, for the systems that were developed underwent successive "taxonomic devaluations": where one had at first glance a genus, it was soon necessary to see an order, and so on.[18]

However, the distinction between generic difference and eidetic difference does not serve as a starting point for any taxonomy in Aristotle. On the contrary, this distinction is closely tied to his theory of opposition and becoming. This is shown by, for example, Chapters 4, 8, and 9 of *Metaphysics* Iota. Aristotle establishes there that the *diafora* has a limit, a maximum extension, and that this maximum difference *(megiste diafora)* is *enantiosis,* a term that we translate "contrariety" (Iota 4.1055a5). The characteristic of a *genos* is that it includes contrary *eide.* And, for this purpose, Aristotle is forced to resystematize more strongly, one might say, his doctrine of opposition. In other texts, in fact, Aristotle distinguishes several forms of opposition *(antithesis)*: relative ones *(pros ti),* the contraries and contrariety *(enantia, tanantia, enantiosis),* possession/privation *(hexis/steresis),* affirmation/negation *(katafasis/apofasis),* according to the presentation in *Categories* 10. But in our passage of the *Metaphysics,* Aristotle is not satisfied with just enumerating these different forms of opposition: he organizes them into a system. Starting from 1055a38, in fact, he not only classes them in order of decreasing "force," but he has them descend from one to the other, as Octave Hamelin has seen clearly:

> In *Metaphysics* Iota 4, this hierarchical classification, whose entire sense can then be perceived, is replaced by the express indication of a

filiation. The most absolute opposition is that of contradictories. That of possession with privation is a determination of the former. Determined in turn, the opposition of possession and privation becomes the opposition of contrariety. Thus, the opposition of relatives must be put in the last rank.[19]

But it is remarkable that it should be precisely in *Metaphysics* Iota, where he gives the most precise exposition of the concepts of *genos, eidos,* and *diafora,* that Aristotle presents the richest and most ordered version of his doctrine of opposition. One may well think that the determination of the field of specific difference led him to a reorganization of the doctrine of opposition. In this deductive and genetic scale of oppositions, the *genos,* which contains contrary *eide,* is placed at a determinate level: above contrariety, it is unable to face an opposition as strong as possession/privation, and it dissolves itself. One thinks here of the criticisms aimed at the Platonists in *Parts of Animals* 1 because they divide according to privation. We now see the real basis of this criticism: like the clumsy butchers of the *Phaedrus,* they do not cut the *gene* according to their real articulations, since they extend the *genos* beyond itself, and thus bring together the *heterogeneous.* And this criticism, coming from the zoologist, has a precise epistemological parallel: the Platonists deny, in fact, the "incommunicability of genera." Thus, to return to the example already used: even though it is correct to divide numbers into even and odd, it is improper to divide beings into those that can and those that cannot be even or odd (for example, on the one hand, numbers, and, on the other, figures). A division of that kind would amount to an attempt to construct a supergeneric science (in this case, arithmetico-geometry). If the *genos* remains, on the contrary, below contrariety and does not surpass the opposition of relation, it cannot actualize all its possibilities, i.e., it remains partially potential. This can be clearly seen in Aristotle's expression in which contrariety is "complete difference" *(teleia diafora)* (*Metaph.* 1055a16). If the differences are complete in this sense, the *genos* is also complete — that is to say, perfect.

The *eidos* is thus indeed a subset of the *genos,* but the *terminus ad quem* of Aristotle's reflections on generic and eidetic differences is not the construction of a taxonomy, but the shaping of the *genos* as a unity of contraries.[20]

At this stage, a question arises that cannot fail to trouble us: has not Aristotle returned to dichotomy by taking it down one step (if I may put it that way) on the scale of forces of opposition, that is, by dividing according to contraries and no longer according to possession/privation? It would appear so, in view of the fact that, after having shown that contrariety is the perfect difference because it is the maximum difference, Aristotle goes on to say, in the *Metaphysics* (Iota 4.1055a19), that nothing can have more than one contrary. Commentators have reasoned that Aristotle thought that the perfect division of the *genos* would be its division into two contraries. Thus, Ross (1924, vol. 2, p. 302) comments on this passage in these terms: "Though Aristotle says that the differentiation is by contraries, he does not in practice confine himself to division by dichotomy; e.g., in *Cat.* 14b37, he divides animals into πτηνόν, πεζόν, ἔνυδον." Thus, for Ross, Aristotle returns here to a dichotomic *method,* which in fact he subverts *in practice.*

Perhaps we can reduce this difficulty if we reconsider what Aristotle means by *enantia.* The usual translation of this term, "contraries," makes it difficult to understand, though I, too, am forced to adopt it. The examples given by Aristotle will lead us to a better interpretation. Let us consider two passages. In *Categories* 11, Aristotle enumerates several pairs of "contraries": good/bad, white/black, sick/well, odd/even, base/noble, courage/cowardice, lack/excess (each of these last two being contrary to right measure), and justice/injustice. In *Metaphysics* Iota 9, he adds male/female and pedestrian/winged. Thus, the attributes called "contraries" are all those that cannot coexist in the same subject in the same relation, but are not "relative to each other" (*Cat.* 10.11b34). Therefore, bad and good are contraries, as are white and black:

> For neither is the good said to be the good of the bad, but one calls
> them contraries, nor is the white said to be the white of the black,
> but one calls them contraries.   (Cat. *10.11b35*)

But double and half are relatives, because the double is said to be the double of the half (*Cat.* 10.11b26). Having thus characterized contraries, we will perhaps be led to understand why Aristotle, while all the time saying that contraries are only two and wanting to divide the *genos* according to contraries, most often divides a *genos* into more than two parts. Let us return to another example given in the *Categories* (11.14a2): defect is the contrary of excess, and the right measure *(mesotes)* is the contrary of both. Aristotle thus indicates that contraries can exist according to several points of view: "for a bad there is as a contrary sometimes a good and sometimes a bad" (*Cat.* 11.14a1). The two contraries defect/excess and defect/right measure do not have exactly the same status, for the first two terms are contraries within bad — which is thus taken as the *genos* within which they display their difference — while the following two are contrary as a bad and a good. This doctrine must have seemed paradoxical to Aristotle's contemporaries, since we read in the *Physics:*

> One should not be troubled by the fact that the same thing may be
> contrary to several others, as movement is simultaneously contrary
> to rest and to the contrary movement; but it is enough to accept this
> thesis that movement is opposed both to rest and to the contrary
> movement, as equality and right measure are opposed to both excess
> and defect.   *(8.7.261b16)*

This leads us to an important point.

I noted above that the *genos* was a "unity of contraries." But this is not a reciprocal relationship: although it is true that every complete *genos* includes contraries, not every contrariety falls within a *genos*. This point is made, with examples, in the same chapter of the *Categories* (14a19): "It is necessary for all the contraries either to be in the same *genos,* or to be in contrary *gene,*

or to be themselves *gene."* And Aristotle gives an example for
each of these three possible cases: white and black are contraries
that are both in the same *genos,* that of color; justice and injustice
are contraries that belong to contrary *gene,* viz., virtue and vice;
and as for good and evil, they are simultaneously contraries and
gene. I note in passing that these examples are very precious,
because they give us a criterion for determining what can be
called a *genos:* a *genos* is a set that can be divided into contraries.
However, the passage just cited shows that this is a necessary
condition but not sufficient: every *genos* is divisible into contraries, but not every pair of contraries is a pair of *eide* that can be
included in a *genos.* Thus, *gene* can be said to be contraries. But,
negatively, we know that no set that is not divisible into contraries can pretend to the name of *genos.*

Justice and injustice, then, are not divisible into contraries
and therefore cannot be called *gene.* In contrast, vice and virtue,
from the fact that they are *gene,* are divisible into contraries: thus
virtue, or rather excellence *(arete),* for the slave would be to obey
well, and for the master, to command well. But, as the *Politics*
tells us, "to be commanded and to command are specifically
different *(eidei diaferei)"* (1.13.1259b37); i.e., according to what
we have established above, they are opposed as contraries within
the same *genos* — so much so that the passage says, a few lines
later, "excellence has within itself a specific difference"
(1260a3). As for vice *(kakia),* it is specified in two contraries:
excess *(hyperbole)* and defect *(elleipsis* according to the *Nicomachean Ethics* 2.5.1106b34; *endeia* according to the passage from
the *Categories* cited above).[21]

Contrariety is thus simultaneously dual and multivocal. But
it can be multivocal in two ways. In the first case, this multivocity of contrariety is not manifest within a single *genos.* Thus, in
the *genos* of bad, the lack does not have any other contrary than
excess; and in this sense, any third term cannot be anything but
an intermediate between defect and excess.[22] But outside the
genos "bad," defect can find another contrary, which is right
measure. Defect and right measure will then be contraries with-

out, however, belonging to the same *genos*. But this multivocity can take on a second aspect when it is deployed within the same *genos*. This is the case with one of the contrarieties found in Book Iota of the *Metaphysics* (9.1058a35), that of pedestrian and winged. For if Aristotle, in this place in the *Metaphysics,* cites only these two terms, it is because they are enough for his purpose. But all his habitual readers know well that this bipartition is an allusion to a tripartition, frequently found in his writings, notoriously in the biological ones, between pedestrian, aquatic, and winged. But these three terms belong to the same *genos,* that of the mode of life (and locomotion) of animals. However, in the contradictory opposition of pedestrian and winged, one cannot find any other contrary of pedestrian than winged by continuing *in the direction of* winged. It is the same for the contrarieties pedestrian/aquatic and winged/aquatic. Thus, Aristotle sometimes cites the three terms of this tripartition together, but sometimes and more frequently two by two.[23] We must therefore conclude from this that one term can have several directions in which one can find a contrary, which is very different from saying that it can have several contraries in the same direction. In any case, Aristotle's biological writings, and notably his work on the *Progression of Animals,* treat foot, wing, and fin as the *eide* of the *genos* "organs of locomotion." But nowhere does he treat one of these *eide* as a sort of intermediary between two others; rather, he considers all three as true *eide*.

But things are still more complicated. From the many complications that we could consider, let us look at this passage from the *Nicomachean Ethics:*

> Now, if continence is good, both the contrary states must be bad, as they actually appear to be; but because the other extreme is seen in few people and seldom, as temperance is thought to be contrary only to self-indulgence, so is continence to incontinence.
>
> *(7.11.1151b28, trans. Ross)*

Everyone knows the first direction in which continence has a contrary: it leads to the pair of contraries continence/incontinence (the latter is defined by Aristotle as an excessive attachment to physical pleasures). But there is a possible second contrary to continence, insensibility to physical pleasures, and that is exactly what this chapter of the *Nicomachean Ethics* discusses. We have difficulty discerning this disposition because it is behaviorally indistinguishable from continence, since both the continent person and the insensible person abstain from bodily pleasures.

Is this an example of a process of dividing the *genos* into *eide?* That seems to be the case, and it may be formalized into a table of virtues and vices, as several interpreters of Aristotle have done. For example, René Gauthier and Jean Yves Jolif (1959, 154) include this chart in their commentary:

Domain	Defect	Mean	Excess
Pleasures and Pains	Insensibility	Temperance	Intemperance

The *genos* divided here into *eide* is the *genos* "attitude toward (physical) pleasures."

First, we see that the process of division of the *genos* into *eide* can be applied at different levels while remaining formally identical, and this within the space of a few chapters — even of a few lines. We first saw division applied to the *genos* "virtue." Now the process of division is displaced and applied to a set with less extension in the same series (in the latter case, frequent in Aristotle, an *eidos* of level one becomes a *genos* in level two). We no longer have, in fact, a division of virtue, but a division in which *one* virtue is included as an *eidos,* and the other *eide* are vices. This variability of level of the *genos/eidos* pair will be very important for our analysis of the biological texts.

Second, the multivocal sequence "insensibility/continence/incontinence," operating *within* a single *genos,* is no more than a particular case of the multivocal sequence studied previously, "lack/mean/excess," which is transgeneric. Thus, a wider or narrower level of generality can, in given instances, combine or suppress the functional validity-conditions of the pair *genos/ eidos*. The possibility of dividing a *genos* into *eide* does not depend on the nature of the beings that it claims to divide, but on the logical conditions that alone make such a division possible. The most important of these conditions is that the *genos* should be divisible into contrary *eide*.[24]

To return to the properties of contrariety, if we look closely at one of the passages of the *Metaphysics* that led me to these analyses, we see that it confirms my interpretation of contrariety as dual in one given direction, but deployable in a multiplicity of directions:

> This being so, it is clear that one thing cannot have more than one contrary (for neither can there be anything more extreme than the extreme, nor can there be more than two extremes for the one interval [*diastema*]).   *(Iota 4.1055a19)* [25]

The use of the term *diastema* suggests a spatial image of the same kind as that which I have used in the translation.

This analysis of the idea of *enantion* shows us, first, how rough the translation "contrary" is and, second, that the division of the *genos* into two contraries in no way leads Aristotle back to Platonic dichotomy.

I must clarify one more point concerning specific difference that will take us even farther from modern taxonomic preoccupations. We have seen that every *genos* is divisible into pairs of contraries, but that not every contrariety is intra-generic. But Aristotle introduces the additional distinction that, among the contraries included in a *genos,* some are relevant in looking for specific difference, while others are not. He deals with this problem at the end of *Metaphysics* Iota 9:

> One might raise the question why woman does not differ from man according to *eidos,* when female and male are contrary and their difference is a contrariety; . . . this question is almost the same as the other, why one contrariety makes things different in species and another does not.
>
> *(Iota 9.1058a29–31, 34–35, trans. Ross, adapted)*

In his examination of the question, Aristotle first eliminates the case of an accidental difference: "There is no difference according to *eidos* between the pale man and the dark man, not even if each of them be denoted by one word" (1058b4). As for sex, it is indeed a "per se difference" (1058a32), "for both 'female' and 'male' belong to animal *qua* animal" (b33). But Aristotle adds another requirement for a difference to be elevated to a specific difference: the contrariety considered must be "in the *logos*" (1058b14 and 18), an expression imperfectly rendered by the translation "in the definition," given by Ross. We will better understand what Aristotle means if we look at what he opposes to this contrariety "in the *logos*." Before introducing this expression, he has just declared that one should not call specific difference a difference produced by the matter, for "matter does not create a difference" (1058b6). But sex is indeed one of the "proper characteristics" of an animal, which means that it is not purely accidental that an animal is male or female, as it is accidental that it is moving or at rest. However, sex does not belong to the specific differences that constitute the *logos* of the animal:

> But male and female, while they are modifications peculiar to "animal," are not so in virtue of its essence but in the matter, i.e., in the body, which is why the same seed becomes female or male by undergoing such or such a modification *(pathos).*   *(1058b21)*

Matter has its role in the determination of the sex of the animal. We know, in fact, that for Aristotle the male semen (a pleonastic expression in any case, since females do not emit semen, contrary to what *History of Animals* 10 claims, which is doubtless the most solid proof of its inauthenticity) as formal cause enforms the female matter. If the semen has enough force, the offspring will be male. To give to sex the status of specific difference is to introduce matter into the definition.

Thus, we return to the definition and to the *ousia,* of which the definition is the formula. In contrast to the division of the sexes, the division of animals into blooded and bloodless is in the *logos* and permits a definition of the essence *(ousia)* of animals

(Parts of Animals 4.5.678a34). This is so much the case that the
doctrine of "proper characteristics" is here only another version
of the doctrine of "per se accidents," which we examined earlier. In fact, the sex of an animal and the fact that the angles of a
triangle equal two right angles are parallel cases: these are both
necessary properties, for it is not by chance but necessarily that
the sex of an animal is determined by the result of the struggle
between the male formal principle and the female material principle. But neither of these two properties enters into the essence
of the defined subject. The primary character of blood or its
analogue, in contrast, makes the division into blooded and
bloodless one of specific differences: it permits the construction
of the definition, the formula of the essence.

Specific difference draws a line of demarcation between substances *(ousiai)* whose logical autonomy it must not fracture.
We find here again the convergence — or more nearly the common root — of the senses that translators have distinguished
(how could they have done otherwise?) in exploding the unity
of the term *eidos:* simultaneously "species" that the specific difference has delineated in the *genos,* the *eidos* is also "form" distinguished from matter, and as such expresses the ontological
basis of the things that it defines, making it also "formal cause."

I must reemphasize another characteristic of the *genos* that
will have a fundamental importance for understanding the criteria of employment of the terms *genos* and *eidos* in zoology. The
pair *genos/eidos* necessarily has a certain classificatory, though
not taxonomic, function because these two concepts also serve as
classes that fit into each other. But the *genos* and the *eidos* (as a
subset of the *genos*) do not have *fixed* positions designating *constant* levels of reality. This can easily be seen by continuing the
quotation of the passage in the *Metaphysics* cited at the end of
Chapter 1:

> There is nothing in the definition except the first-named *genos* and
> the specific differences; the other *gene* are constituted by the first
> *genos* together with the specific differences that are taken with it,
> e.g., the first *genos* may be "animal," the next "two-footed animal,"

> and again "two-footed and featherless animal," and similarly if the
> definition includes more terms.   *(Zeta 12.1037b29)*

The qualifier "first" does not here imply any preeminence, but
only indicates a relative position in the process of growing specification. The *genos* therefore can no more function as "genus" in
the taxonomical sense than can *eidos* as "species." For taxonomists, obviously, the essential characteristic of such concepts is
that they must designate a constant level in the scale of beings.
For Aristotle, in contrast, one may say that the concept of *genos* is
classificatory but not taxonomic: as the term does not designate a
fixed level of classification, to say of a collection of objects that it
makes up a *genos* is not, from a classificatory point of view, to say
anything more than that it is subdivided into *eide.* Similarly, to
say of a being that it is an *eidos* is to say nothing more than that it
has been carved out of a *genos* by slicing according to a specific
difference, and says nothing at all about its absolute degree of
generality. The pair *genos/eidos* thus constitutes a diairetical tool
functioning at any level of generality at all.

However, Aristotle apparently established two limits, upper
and lower, beyond which the pair *genos/eidos* cannot go in its
multilevel operation. The lower limit is the *atoma,* which, as
their name indicates, cannot be divided further. Determining
the upper limit is more difficult; the "traditional" interpretation
of Aristotle would situate this upper limit at the level of the
categories, which are *"gene* of beings." [26] That was already the
interpretation of Porphyry:

> In each category there are most general genera *(genikotata)* and, on
> the other hand, most specific species *(eidikotata),* and others that are
> intermediate between the most general genera and the most specific
> species. A most general genus is that above which there is no other
> genus in the rising series; a most specific species is that after which
> there is no species in the descending series. [27]

On the other hand, as the *genos* is a unity of contraries according
to the *ousia,* the relevance at several levels of the diairetic pair
genos/eidos corresponds to the relevance at several levels of *ousia*

itself. Here I shall only note this convergence, without deciding which of these two variabilities of level is the cause of the other. In the context of the variability of the meaning of *ousia,* one understands best how much all Aristotle's concepts may take on a variety of meanings, in that they may be applied not only in several directions, but also, as it were, in varying degrees. Thus, in the case of *ousia,* Aristotle gives it several characteristic properties, of which one example is that of being "separate" (see *Metaph.* Zeta 3.1029a28). This leads him sometimes to deny that matter is *ousia,* as he does in that same Zeta 3, and sometimes to support the contrary thesis, as he does in Eta 1.1042a32 ("that matter is also *ousia* is obvious"). Simplifying, perhaps excessively, the subtlety of Aristotle's analysis, I believe that we have here a variation of degree: matter is less *ousia* than is, for example, the composite of matter and form. Or, more exactly, although it is *ousia,* matter draws less than the concrete composite from the logical content of the notion of *ousia.* This distinction is, in my eyes, one of the keys to Aristotle's text: it will be very useful to us when we approach the use of *genos* and *eidos* in the biological corpus.

However, behind an immediate resemblance, there is a fundamental difference between the variability of *ousia* and that of the pair *genos/eidos.* For though the latter is a universal schema able to function in very different domains and levels, the variations of *ousia* are all ordered in relation to a fundamental privileged level. In the biological domain, there is no doubt that for Aristotle *ousia par excellence* is the individual concrete animal: this horse here. But even in biology, Aristotle runs up against the epistemological difficulty that his theory attributes full and entire reality to concrete individuals but still maintains that there is no knowledge except of the universal. Scientific discourse thus will range over what the *Categories* calls "second *ousiai*" (5.2a14). That is where the variability of level that I have mentioned is involved. Thus man, taken as a species of animal, is an *ousia,* which implies that there is a definition that, as the *Topics* tells us, "should be true of every man" (6.1.139a26), which means that man has a *ti en einai,* since the *Topics* definition is presented as signifying the *ti en einai* (1.4.101b21, 38, et al.; cf. *Metaph.* Zeta 4.1030a6: there is no *ti en einai* except of things that have a definition). That is probably what an expression like this one, from the *Parts of Animals* (1.1.640a34), means: *"tout' ēn to anthrōpōn einai."* But Aristotle also speaks of a *ti en einai* and an *ousia* of "bird," which is not an animal species, as man is, but a larger class. Thus, he writes that the fact of having wings enters into the *ousia* of "bird" (*PA* 4.12.693b13). But the character of *ousia* or of *ti en einai* can be attributed to much larger classes. Note this very interesting passage, which takes up the example of "bird" again:

> In short, the lung is for respiration; it is lacking blood and as such it is adapted to a certain class *(genos)* of animals, but one that has not received a common name, in contrast to the *genos* "bird," which has received a name as such. So that just as the essence *(to einai)* of bird presupposes a certain condition, so too for these animals the fact of possessing a lung belongs to their *ousia.* *(3.6.669b8)*

Not only does Aristotle situate *ousia* at several levels here, but he also seems to be putting the reader on guard against common linguistic practice, which can hide the *ousia* of a family of animals: although they have not been named, the group of animals with lungs is just as much an *ousia* as the family of birds, which are also characterized by a property, doubtless that of having wings. It is possible to ascend even higher on the taxonomic ladder, and we shall see in the next chapter that there are many passages where Aristotle defines "animal" or even "living thing" by characteristic properties (nutrition, sensation, movement). Even classes that general ought to be considered as *ousiai.*

Nevertheless, as I suggested above, there is a way of going from one *genos* to another, permitting the closure of each *genos* on itself to be surmounted — and that is the way of analogy. I would like to say a few words about this notion, since it has a special importance in biology. Pierre Aubenque (1978) claimed, quite rightly, that he should denounce a bit of centuries-old nonsense that attributes to Aristotle a doctrine called "analogia ad unum." I must strongly second Aubenque by maintaining that the only analogy found in Aristotle is analogy "of proportion," doubtless of mathematical origin, which the *Nicomachean Ethics* defines as "an equality of relation between at least four terms" (5.3.1131a31).[28]

Above the generic gap, analogy can establish relationships between *heterogeneous* beings, permitting propositions of the type: "that which is (a) in *genos* A is (b) in *genos* B." For example:

> We cannot say that all beings have the same elements and the same principles, except by analogy, as if one were to say that the principles are three in number—form, privation, and matter—but each of them is other in each *genos*; for example, in [the *genos*] color they are respectively white, black, and surface; in the *genos* of day and night they are light, darkness, and air.   (Metaph. *Lambda 4.1070b17*)

White is thus to color what light is to day, and so on.

To finish this logical examination of the concept of *genos* (a selective examination to the extent that it concentrates on those aspects that are significant for our study of Aristotelian zoology), a comparison with its parent etymological notion, *genesis,* seems enlightening for our later purposes. In fact, generation occurs from one contrary to another (e.g., *GC* 2.4.331a14; *Metaph.* Iota 4.1055a8); and to the extent that it is a "movement according to substance," it is distinguished from simple alteration (in which the being changes only according to its affections).[29] Thus the *genos,* a collection of *eide* that are contrary according to the *ousia* and not according to the "proper affections," would be (so to speak) a fixed generation corresponding, in logical space, to what *genesis* orders in a chronological succession.

In the last pages of the treatise *On Generation and Corruption,* Aristotle explains that the survival of living things through their descendants, by means of generation, is an imitation of the permanence of eternal beings by beings that cannot attain individual eternity. At the beginning of *Generation of Animals* 2, he returns to this idea, associating it with the notion of *genos:*

Since the nature of this *genos* [i.e., the *genos* of beings subject to generation, in opposition to eternal beings] is unable to be eternal, that which is born is eternal in the manner that is open to it. Now, it is impossible for it to be so *numerically;* . . . it is, however, open to it to be so *specifically.* That is why there is always a *genos* of men, of animals, of plants.   *(2.1.731b31)*

Notice in this passage that *genos* successively designates two groups with very different extensions, since in the first case it is all beings subject to becoming, and then right afterwards it is men, other animals, and plants, and thus is presented as the territory of *genesis.*[30] My analysis thus restores to *genos* its sense of "lineage," based on the very etymology of the word; this is also its fundamental sense in Aristotle, as is shown by the first definition of *genos* given in *Metaphysics* Delta, in the chapter devoted to this concept: "We speak of *genos,* in the first place, when there is an uninterrupted generation of beings having the same *eidos*" (Delta 28.1024a29). I have shown elsewhere[31] that the interpretation of this passage is situated between two extremes: this *eidos* is, at a minimum, the common "look" that the members of the same *genos* conserve from one generation to the next; but it can also be (and I think this more probable) the "form" contained in the sperm of the male founder of the lineage. I should say *patri*lineage, since females do not, for Aristotle, have a truly *gene*tic role, as I have explained at length in the article cited above. And, in fact, a lineage can receive its denomination from the maternal side only in quite exceptional cases, and in a derivative sense: "The *genos* is rather named after the generator than after the matter, although *genos* can sometimes receive its name from a female, as in the case of the descendants of Pyrrha" (*Metaph.* Delta 28.1024a34). This genetic and lineage sense is also to be found in several places in the *Politics.*

When we see Aristotle using, one might say entirely naturally, the term *genos* to designate what we would call an animal species, or even a variety, we do not say that he has thereby broken with the logical usage of the concept, and we do not blame him for a lack of coherence and rigor, as do the commentators. Since the term *genos* is not limited to the designation of a determinate level of reality, because it is not a taxonomic concept in biology, it can, as we have determined from his logic, designate the widest branchings as well as the smallest variety of animals or plants. Ultimately, the latter case is nothing more than a limiting use of *genos* as the unity of contraries: throughout all the influences that play upon the *genos,* from its appearance to its disappearance, *genos* maintains the identity of the living form and distinguishes it from all other living things.

Porphyry, too, was sensitive to this relationship between *genos* and generation:

> Neither genus nor species appears to have one sense, for genus can mean (1) a collection of things related to one another because each is related to some one thing in a particular way. In this sense, the Heraclids are said to be a *genos* because of the relationship of descent from one man, Heracles. The many people related to each other because of this kinship deriving from Heracles are called the *genos* of the Heraclids, since they as a *genos* are distinct from other *gene*.
>
> In another sense, *genos* means (2) the source of each man's birth, whether from his father or from the place in which he was born. When we say that Orestes is of the *genos* of Tantalus, . . . that Pindar is of the Theban *genos* . . . This meaning seems to be obvious. . . .
>
> In another sense, *genos* means (3) that to which the *eidos* is subordinate. This sense is taken to define *genos,* perhaps because of the similarity of this sense with the two former senses. For *genos* in this sense is, in a way, a source of the *eide* under it, and it also seems to contain the whole subordinate multitude.
>
> (Isagoge, *Busse 1.8; trans. Warren [1975], with minor changes)*

The use of the terms *genos* and *eidos* in the zoological domain has sorely tried the perspicacity of the interpreters. To avoid a complete review of all interpretations, let me begin with one of the best of them, if not the very best: the solid and detailed study that David Balme devoted to these two terms in his 1962 article, "*Genos* and *Eidos* in Aristotle's Biology."

With Balme, we must recognize and resist the stubborn attempt of previous commentators to make the pair *genos/eidos* function as the pair "genus/species" later would. It will be enough to call attention to one group of interpreters who all, without exception, do this. The group I have in mind try to construct great synthetic interpretations, but in fact neglect Aristotle's biological works (close to a third of the Corpus Aristotelicum as we have it) and mention them only to fulfill their promise of a complete exposition of Aristotle's doctrines. For example, Léon Robin (1944, 175–81), in the more than three hundred pages of his *Aristote,* devotes at most seven to "Aristotle's biology," with a few allusions thrown in elsewhere. Better prepared to study other aspects of Aristotle's work, these interpreters were alienated by their education from the problems that preoccupy us here. "Those of our contemporaries who know proportionately as much zoology and biology as Aristotle knew are not professors of philosophy," Gilson acidly comments (1971, 11).

The interpreters who deserve Balme's attention, and ours, are those who have looked seriously at the biological texts, to such an extent that they could not avoid noticing the textual evidence of a usage of the terms *genos* and *eidos* that is aberrant from a taxonomic point of view. Nevertheless, they maintain, against that evidence, that Aristotle had a taxonomical project based on these concepts. But even the most sympathetic reader, despite the subtlety of some interpretations, could not maintain that *genos* and *eidos* are even distant prefigurations of the "genus" and "species" of the later taxonomists. And Balme, not without obvious satisfaction at the despair of his predecessors, cannot resist pointing out the inextricable confusion to which the traditional understanding leads. Let us summarize the gist of Balme's argument, without necessarily following his order of exposition or using his examples.

On the one side, there is no dearth of texts in which the two terms seem to have, at least approximately, the significance of "genus" and "species" in the modern sense. Thus, the *History of Animals* opens with methodological considerations that seem to be consonant with taxonomy:

> With regard to animals, some have all their parts mutually identical, some have parts of a different character. Some parts are identical in respect to the *eidos;* for example, one man's nose and eye are identical with another's nose and eye, one's flesh with another's flesh, one's bone with another's bone; and the same applies to the parts of a horse, and of such other animals as we consider to be specifically identical; for as the whole is to the whole, so every part is to every part. In other cases, they are, it is true, identical, but they differ in respect of excess and defect: this applies to those whose genus is the same; and by genus I mean, for example, bird and fish: each of these exhibits difference with respect to genus, and of course there are numerous species both of fishes and of birds.   *(1.1.486a14)*

Obviously, I have left the translations of *genos* and *eidos* (except the first use of the latter term) as "genus" and "species." This comparison of animals according to their "parts" does seem to permit grouping them into "species" that are themselves grouped in "genera," and thus seems to contain a potential natural history, and the examples given would seem to confirm that impression. One might even dream of the method of Adanson, who (in botany) wanted to construct a natural classification by comparing *all* the "parts" of plants. And Aristotle pushes his construction farther, since he writes a few lines later:

> Some animals, however, have parts which are not specifically identical, nor differing merely by excess and defect: these parts correspond only "by analogy," of which an example is the correspondence between bone and fishspine, nail and hoof, hand and claw, feather and scale: in a fish, the scale is the corresponding thing to a feather in a bird.   *(486b17, trans. Peck)*

It appears in this passage—and there are many like it in Aristotle—that he is trying to give himself a way of including all animals in his picture, no matter what their immediate differences. In any case, these are the texts that seem, in the words of Pierre Louis already cited, "to open the way to comparative anatomy."

But on the other side, even more numerous examples can be found against the rule of reading *genos* and *eidos* as "genus" and "species." Thus, Balme shows, particularly against Bonitz, that the term *genos* designates ultimate species and even varieties of a given species; and Balme cites the dog, which is called a *genos* in the *Parts of Animals* (2.14.658a29), whereas in the *History of Animals* (6.20.574a16) Aristotle says that "there are several *gene* of dogs." Balme then gives a long list, almost exhaustive as far as I can tell, of the cases in which *genos* is used to designate what we call species (pp. 85–86). The term *genos* thus does not always have the degree of generality that attaches to our term "genus." Let me give an example, not cited by Balme, which should add even more to the perplexity of the taxonomical interpreters. In the *Generation of Animals* (2.1.732b23), Aristotle cites as a *genos* the family of snakes minus the viper. In this case, modern taxonomists themselves do not have a word for designating such a group.

A second difficulty is that some animals are ranked sometimes under the term *genos* and sometimes under the term *eidos.* Balme cites five cases of this kind: cicadas, sea lice, jackdaws, sea urchins, and caterpillars (p. 86).

Finally, Aristotle sometimes uses the word *eidos* to designate classes that, even if they do not have locations in the *Systema naturae* of classical naturalists, have a degree of generality well beyond not only our species but even our genera and our families. Thus, in the *Generation of Animals* (1.11.719a7), *eide* designates no less than the classes of vivipara and ovipara.

Faced by these two groups of texts, it would not be difficult to conclude that there is simply a contradiction between them, or at least a distortion, and that Aristotle was ultimately unable to carry out his project of classification. It would, in any case, be interesting to list and to analyze the ways that commentators have juggled, reduced, or justified this textual distortion, but that would take us away from our present topic. I may, however, remark here that most of these "explanations" have as a common point of departure the positing of the problem in terms of the lack of fit between Aristotle's theoretical project and the language (terms) that he uses to state it and to carry it out. Thus, interpreters have extensively used — and abused — the distinction noted above between the "technical" and "nontechnical" senses of terms. So that no ambiguity remains on this point, especially following what I said earlier, let me make clear that this distinction is not, in my opinion, totally meaningless, for Aristotle did indeed take terms from current language on which he conferred technical senses — as do all philosophers. And it would be at least surprising if, after "technicizing" terms, Aristotle gave up speaking the language of his contemporaries. Thus, we find, for example, the word *diafora* used in the current sense of "disagreement, divergence" in the ethical and political texts.[32] The use of this distinction for explaining contradictions that we believe we have found in the texts should be totally avoided. Balme recognizes this: "That the same words should have sometimes a technical and sometimes a non-technical use, is not necessarily significant" (p. 83). Anyway, in the case before us we have far too many supposedly aberrant texts for that to be a reasonable explanatory procedure.

To complete our consideration of the attitude of commentators faced by this immediate contradiction in the uses of *genos* and *eidos* in Aristotle's biological texts, let us look at a remark by Pierre Louis in his edition of the *History of Animals*, which reveals the impasse of the interpreters who want to make a taxonomist of Aristotle, and which also contains a hint of one of the ways to get out of that impasse. The remark comes in a note to the passage (2.14.505b10) in which Aristotle, after having announced that he is going to deal with the *genos* of serpents, writes that there are several *gene* of water serpents. Louis comments: "*Genos* is sometimes a subdivision of *eidos* and signifies variety, race";[33] and he gives two other references relating to cattle and to hares. Louis's position is not incoherent: if Aristotle is really a taxonomist, the *genos*/genus is above its *eide*/species, and the species sea serpent is included in the genus serpents. In this passage, the word *eidos* itself does not appear, but for Louis it can be assumed, since in talking about sea serpents Aristotle determines in the scale of the classes that he is thought to have developed, the *level* of the species. But we have to realize that the term *genos* is also used *below* the level of the species. Thus, in order not to call into question the idea that there is a taxonomic usage of the words *genos* and *eidos* in Aristotle, Louis enunciates a proposition that should horrify every Aristotelian: that an *eidos* can be divided into *gene*. However, there is a sense in which one could understand what Louis says as an invitation not to insist on keeping *genos* at a given level of generality; but in Louis that interpretation is only "potential," unsuspected by him.

We must now look at the solutions that Balme proposes for escaping the textual difficulties related to the use of *genos* and *eidos*. In the first place, Balme recognizes, and the point must be accepted once and for all since his article, that *genos* and *eidos* are not taxonomic concepts:

> The above evidence shows, wherever it is possible to test the deno-
> tations of γένος and εἶδος by reference to actual animals, Aristotle's
> usage makes no taxonomic distinction between them. Γένος means
> a "kind" at any level from the most abstract group to the immedi-
> ately visible type. Εἶδος is far less commonly used, and represents no
> group that γένος does not also represent. If they differ in meaning,
> the difference is not that of higher and lower rungs on the same
> ladder. They belong to different ladders, and the original difference
> sometimes shows through, γένος being a statement about kinship
> and εἶδος a statement about shape or form.   *(P. 87)*

Thus, Balme relies on an analysis of terms based on etymology. I do not deny the relevance of etymology, but it cannot by itself resolve the problems posed by the use of the terms *genos* and *eidos* in the biological corpus — particularly, but not only, when they designate animal classes. Thus, for example, it is totally impossible to pretend that when Aristotle writes, only a few lines apart in the *History of Animals,* that, on the one hand, there are two *gene* of ouzel, one black and the other very white — or just white, depending on the manuscript (9.19.617a11) — and that, on the other hand, there are three *eide* of thrush, the ixoborus, the trichas, and the iliades (9.20.617a18), the use of different terms can only be explained by an allusion to kinship in the first case and by a reference to the forms of the animals considered in the second.

But the heart of Balme's response to the textual challenge lies elsewhere. In fact, when he places side by side two families of texts, one in which the terms *genos* and *eidos* seem to have a real taxonomic value, and the other in which their use seems totally unregulated, Balme cannot help thinking, like other commentators, that there is indeed a contradiction in Aristotle. What distinguishes Balme from his predecessors, and what is in my opinion an enormous step forward, is that he does not locate the contradiction between the logical books and the biological books, but within the biological corpus itself. The step is enormous because Balme thus posits the conceptual unity of Aristotle's thought. In this sense, we all start from Balme.

In fact, Balme has the same first impression as that with which I began the second part of this chapter: namely, that there seem to be two groups of texts in respect to the use of *genos* and *eidos* in the biological corpus. In the first group, that in which *genos* and *eidos* function as concepts that we might call "paleo-taxonomic," Balme counts seven passages — plus two doubtful ones — but only by amputating from the biological corpus *Parts of Animals* 1, on the pretext that it is methodological. These passages are: *HA* 486a16–b21, 488b30–32, (490b7–491a4?), 491a18–19, 497b9–12, (505b26–32?), 539a28–30, *GA* 784b21–23, and *Long. Vit.* 465a4–7. I call these passages "paleo-taxonomic" because we might extract from them a common classificatory doctrine according to which one would use the word *genos* for a family of animals having between them differences of degree, whereas animals with a similarity in all

their parts would belong to the same *eidos*. I quoted the first of
these passages above (p. 76). Balme limits his count to passages
that are more or less programmatic; we could enlarge this first
family of passages not only with several passages from *Parts of
Animals* 1 (e.g., 1.4.644a16), but also with other passages from
the biological corpus that appear to assume this classificatory
doctrine as established, in that they *allude* to it. Thus, at *PA*
4.12.692b3:

> The difference that exists between birds is that of the excess and
> defect of the parts, or a difference of the more and the less. There are
> among them those that have long legs, others that have short legs;
> some have a wide tongue, others narrow.

The rest of this passage constantly uses analogy, without men-
tioning the word: for example, feathers, beaks, and so on are to
birds what scales, teeth, hands, and trunks are to other animals.

In the second group of passages, in contrast, Aristotle seems
to waste the rigor of the logical schema that he had developed in
the first group of passages. I have already given many examples.
Aristotle's biology would thus carry the traits of an incomplete
project: that of applying to the living world the concepts of *genos*
and *eidos* as defined in the logico-metaphysical writings. The
passages of the first group would then be, in a way, the spearhead
of that invasion of biology by logic. Thus, the seven (or nine)
passages mentioned by Balme appear to be out of place: "They
share certain features, in which they disagree with the rest of the
biological works but agree with *PA* 1 and to some extent with
the logical works" (pp. 96–97). According to Balme, then,
there is a contradiction within the biological corpus itself, since
it offers us, as it stands, the image of an incomplete project:

> If that is so, it would seem that, contrary to what is often said,
> Aristotle must have developed the technical distinction from logic
> and not from biology. [Balme has a note here that he disagrees on
> this point with LeBlond.] He must have intended to apply it to
> biology, for it is incredible that he should have abandoned system-
> atics in the very field where it has proved most fruitful, and
> proved so for the very reason that he himself predicted in *An. Post.*

2.13–14. In that case his biological work is incomplete. This is indeed self-evident: what is missing is a straight descriptive zoology, together with a classification system that a descriptive zoology needs if only for orderliness.   *(P. 98)*

This part of Balme's article shows that his difficulties had the same origin as those of the other commentators: like them, he had not liberated himself from the idea that Aristotle must necessarily have had a taxonomic goal. This is the idea that ultimately made Balme, or at least the Balme of 1962,[34] fall back into the errors of his predecessors, which he had so correctly criticized. Before showing that there definitely is no fixed point in Aristotle's biology to which one might anchor an animal taxonomy, even as a project, I would like to prove that Balme's interpretation should be rejected even by someone who accepts the split it introduces — in my opinion, improperly — into the biological corpus. In fact, even in the first group of passages — composed, according to Balme, of seven (or nine) texts in which "the technical distinction between genus and species appears obligatory" (p. 96) — *genos* and *eidos* are notions that wander too much to supply a stable basis for an animal taxonomy.

In the last passage of the list given by Balme, we read:

I say that things differ according to *genos* when they differ as man and horse (for the *genos* of men has a life longer than that of horses), but according to *eidos* as man differs from man.

(Long. Vit. *1.465a4*)

While in this passage man and horse are two *gene,* and two horses differ according to *eidos,* the other passages in Balme's list support a different viewpoint, closer in that respect to most of the logical texts. For example, in *HA* 1.1.486a19, Aristotle writes of horses: "We say that they are the same with respect to *eidos*." Even these passages, although programmatic in that they give definitions ("I say," "we say"), are not in agreement about the level of generality at which the concepts of *genos* and *eidos* should be placed. From an examination of the passages, and from Balme's analyses, we can and ought to draw the conclusion that neither *genos* nor *eidos,* nor the pair *genos/eidos,* defines a fixed level of generality, and that consequently, although they function often enough as classificatory concepts, they are never taxonomic concepts.

This semantic slipperiness and this apparently irregular usage of the notions of *genos* and *eidos* as taxonomic tools should not surprise us at this stage of our study. What is really surprising is that interpreters should have sought in *genos* and *eidos* a solid foundation for an eventual Aristotelian taxonomic project: the logico-metaphysical works, in fact, as we have seen, make the variability of level one of the most obvious properties of these two concepts, particularly when they function together. This stubbornness of the commentators appears to me to have principally historical and etymological causes: the passage of the pair *genos/eidos* to the Latin pair *genus/species* and all the equivalent pairs in the various European languages has created in the commentators a desperate desire to see fixed classes in Aristotle's *genos* and *eidos.*

But once the untaxonomic character of *genos* and *eidos* has been realized from the texts, are there no other notions that could, by setting a constant level of generality, serve as a fulcral point for a taxonomy, at least in prospect? It seems to me that this task has been foisted off on the concept of "the greatest *genos*" and on the doctrine that opposes differences of degree and analogical difference. We must therefore rapidly examine that concept and that doctrine.

In fact, considered in itself, the expression "the greatest *(megista* or *megala) gene*" seems incapable of designating a fixed level of generality, since the adjective *megas,* whether or not used in the superlative, only qualifies the word *genos,* and it would thus be grammatically logical that the expression *megiston genos* should inherit the same qualities as the word *genos,* and in particular the same variability. Nevertheless, some interpreters have seen in the "greatest *gene*" a fixed level of generality.

Thus, Pierre Louis writes with admirable confidence: "In

Aristotle's classification of animals, the expression τὰ γένη μέγιστα (or μέγαλα) has a precise sense: it designates the great divisions of the animal kingdom, the classes." [35] A comparison of three passages will show the road to the truth:

> Here are the greatest families *(gene megista)* into which animals have been divided: one is that of birds, another that of fish, another that of testaceans, . . . another that of crustaceans, . . . another that of molluscs, . . . another that of insects. . . . None of the other animals are in great families *(gene megala).* (HA *1.6.490b7)*

> For in the group of animals, one may distinguish these two greatest families: that of blooded and that of bloodless.
>
> (HA *2.15.505b26)*

> There are four greatest families *(gene megista)* [of crustaceans]: they are called spiny lobsters, lobsters, crayfish, and crabs. Each of these families has many *eide,* which differ not only in form but especially in size.   (PA *4.8.683b26)*

In these three passages, the *megista gene* are indeed involved in a fitting together of more or less large animal classes. Thus, the notion is indeed classificatory, but it is not taxonomic, because, far from referring to a fixed level of generality, it designates in each of these three passages a different level of generality. I thought at first (and I wrote in the first French edition of this book) that the expression *megista gene* meant a group large enough to be subdivided several times. James Lennox (1980, 338), in an article to which we shall have occasion to return, proposes a related interpretation, but concerning the intension of the concept of *megiston genos* rather than its extension: "Aristotle's zoological μέγιστα γένη are distinguished from one another *not* by one trait that is necessary and sufficient for membership in a kind, but by an organized set of general traits. Birds are oviparous, feathered, beaked, two-legged, and winged. Fish are cold-blooded, gilled, scaled, aquatic, oviparous, and so on." I now think, particularly in view of the third of the passages that I have quoted, that the expression *megista gene* designates quite simply "important *gene,*" in that they have many *eide,* and that the superlative *megista* is not meant to signify anything other than this profusion of *eide.* When Aristotle considers these classes as a whole and does not want to insist upon the extent of their divisibility, he normally designates them with the word *genos:* for example, for the crustaceans, of which he says several times that they form one of the *gene* of bloodless animals (*HA* 4.1.523b1; *PA* 4.5.678a30; *GA* 1.14.720b4).

But what about analogical difference and difference of degree? We find in many places in the biological corpus a doctrine according to which within the *genos* there is found only a difference of degree, while between *gene* there is an analogical difference. Thus, in *PA* 1.4.644a16:

> For those of the families *(gene)* that differ in degree, i.e., according to the more and the less, have been arranged under one *genos,* while those that have an analogical difference have been put in separate genera. I mean, for example, that a bird differs from another bird by the more and the less, that is, by degree (one has the wings longer, the other shorter), while a fish differs from a bird by analogy (for what is feather in the one is scale in the other).

Let us begin with the question of the difference of degree, relying heavily on Lennox's remarkable article, which shows very well the difficulties in reconciling the logical doctrine of division of a *genos* into *eide,* and the thesis, apparently peculiar to the biological corpus, according to which within a *genos* the *eide* differ according to the more and the less.[36] Lennox thinks that these two different descriptions of the unity of the *genos* across its differences are reconcilable, particularly by deriving the doctrine of the difference of degree from another doctrine asserted by Aristotle, that the *genos* is the matter for the differences that operate within it,[37] and then by resorting to teleology to establish the unity of each animal species:

> Differentia by differentia, compared out of context, the species of a genus differ only by "excess and defect," or "the more and the less." But when the balanced unity of those differentiae in each species is studied, and when it becomes clear that the possession of just this degree of length of beak is needed for the members of this species to get their food, Aristotle insists that the forms of a genus are indeed different in form, and not merely in degree. *(P. 344)*

Let me make a few remarks concerning the questions of special interest to us here. Starting from the texts, we can posit the following distinction: among the *gene,* which, by definition, are all divisible into *eide,* some admit variations of degree at the level of their *eide,* others do not. In the latter case, there may even be an opposition between difference of degree and specific difference. Lennox quotes on this subject a passage from the *Politics* (1.13.1259b36), in which Aristotle says that to command and to be commanded differ according to *eide* and not according to the more and the less. This passage is not unique. At the beginning of the same book of the *Politics* (1.1.1252a9), Aristotle writes, probably against the Platonists, that there is between the statesman, the king, the head of a family, and the master of slaves a difference in *eidos* and not a difference of more and less, according to the number of people over whom one's power is wielded. In contrast, the "physical characteristics" of the "parts" of animals accept variation of degree within a *genos:*

> [In each *genos*] the parts differ, not in having an analogical relation (as in man and fish the relation of bone and fishspine), but rather by the physical characteristics, such as largeness and smallness, softness and hardness, smooth and rough, and properties of that kind; in short, they differ by the more and the less.   (PA *1.4.644b11*)

Lennox is right to put these biological passages into relation with, among others, the passages in the *Categories* concerning the more and the less (e.g., 8.10b26). But I would go even farther than Lennox in noting that the *Categories* says that certain contraries accept the more and the less and that others do not accept it — or, more exactly, that some things conjointly accept contrariety *and* the more and the less. This is so for many qualities (8.10b12 – 11a14) and for action and passion (9.11b1 – 2). This should be put into relation with the distinction that I have made above, according to which, among the contraries, some accept intermediaries between them and others do not. Thus, there are cases in which one goes from one contrary to the other by a continual variation of degree, which does not mean that the extremes of that movement are not true contraries. For example, in becoming better and better little by little, a man can go from vice to virtue, but nevertheless vice and virtue remain true contraries, and one cannot say that they differ from each other according to the more and the less (see *Cat.* 10.13a22). Thus, difference of degree within a *genos* is not absolutely incompatible with the contrariety of *eide* within the same *genos,* a contrariety that we have seen is a fundamental property of the pair *genos/eidos.* The doctrine of the difference of degree, although nearly peculiar to the biological corpus, cannot claim to substitute in this field for the rules that regulate the use of the *genos/ eidos* pair in the logical works. In brief, the doctrine of the more and the less does not introduce a conceptual break between the logico-metaphysical books and the biological books.

Even if we do not have a passage that tells us so explicitly, the reason Aristotle uses this concept of difference of degree in biology seems obvious, in view of the clear advantages he can get from it both for the description and for the ordering of the parts of animals. But what is the epistemological status of such a doctrine? I think the answer can be found in the passage in the *Parts of Animals* quoted above (644b11), at least if we notice that it follows another passage, already quoted, in which Aristotle asserts that we should accept the traditional distinctions "when they are correct" (644b2). Aristotle thus seems to take this doctrine from a certain tradition that opposes difference of degree and analogical difference, even though he has attached it to his own thought by applying to it an Aristotelian terminology. It is difficult to say who "the people" (644b3) were from whom Aristotle borrowed his distinctions, and particularly to what degree they had a scientific or technical "specialization." So this doctrine of the more and the less seems to me to be an empirical procedure, not developed by Aristotle himself; he uses it because it provides practical advantages without casting doubt on the properties of *genos* and *eidos* as defined by the logico-metaphysical books. This doctrine therefore cannot furnish the basis for an animal taxonomy: that will appear still more clearly when we attack the problem of analogical difference.

It is perfectly consonant with the logical properties of analogy that the difference between *gene* in biology should be analogical. There is even a terminological identity: in both domains, Aristotle uses the word *analogia,* especially in the expression *kat analogian,* and the neuter *analogon.* In fact, I can immediately say about analogy what I said about the "greatest *gene*": the fact that it is defined in relation to the concept of *genos* ought to be enough to convince us that the analogical relation does not designate a fixed level of generality. But I would like to show that more exactly, supported by texts.

The commentators who have tried to give a taxonomic function to the analogical relation seem to suppose that the analogy between feather and scale, for example, fixes the *genos* at the level "bird" and "fish." In other words, they figure that because feather and scale are said to be analogues, therefore birds and fish are *gene.* But that is absolutely anti-Aristotelian. A logical examination of the concepts of *genos* and *analogia* shows us, on the contrary, that feather and scale can be said to be analogues *from the moment* one decides to take "bird" and "fish" as *gene.* But if the level called *genos* changes, which as we have seen can occur by definition, analogy, too, changes level. This can be shown by the comparison of two passages, pointed out by Balme, placed a few pages apart in the *Parts of Animals:*

> The animals that do not have [bone] have something analogical; in fish, for example, in some there are fish spine and in others cartilage.   *(2.8.653b35)*

> The nature of cartilage is the same as that of bone, but they differ according to the more and the less.
>
> *(2.9.655a33; cf. HA 3.8.516b31)*

To ask whether there is between bone and cartilage an analogical difference or a difference of degree is simply not an Aristotelian question. In the first passage, we have two *gene,* bony animals and cartilaginous animals, which have between them an analogical relation. The point of view of the second passage is not the same. The chapter in which it is found studies the nature and functions of the skeleton: bone and cartilage are now two different "species" of material used by nature as "support" for the body. (At 654b29, Aristotle writes that nature operates as sculptors do when they work with a soft material like clay, which obliges them first to construct a rigid armature.) Within the *genos* constituted of "parts" assuring the rigidity necessary for the animal, there are variations of degree, particularly of size and hardness. The difference between bone and cartilage relates, among other things, to hardness: their difference is thus a difference of the more and the less.

One may find other examples of these changes in perspective that have the effect of shifting the relation of analogy. Let us look at an example very close to the preceding one, comparing two passages:

> In some animals, the parts are not of the same *eidos* and do not differ by excess or defect, but differ by analogy: that is the case, for example, of bone in relation to fish spine, nail in relation to hoof.
>
> (HA *1.1.486b17*)

> There exist certain parts that are, to the touch, related to bone; for example, the nails, hoofs, claws, horns, bird beaks. All these parts are possessed by animals for the sake of their defense.
>
> (PA *2.9.655b2*)

The first passage claims that what nail is for *genos* A, hoof is for *genos* B: there is an analogical relation comparable to that which says that feather is to the bird what scale is to the fish. In the second passage, in contrast, nail and hoof are considered as *eide* of the *genos* "organs of defense," even if Aristotle does not say so explicitly. From this point of view, nail and hoof are no longer analogues. One may even suppose, when Aristotle notes a little later (655b11) that all these parts are composed of earth, that this is a way of stating their unity as a *genos*.

In the light of what I have just said, there is a passage that becomes particularly interesting; although it is found in the *Posterior Analytics* (2.14), it relies on a biological example. What is at issue here is choosing the divisions in order to obtain relevant groupings to which one may attribute properties. Some of these groups have a name — for example, the class of birds; others do not — for example, the class of animals with horns. Aristotle says at the end of the chapter:

> There is another way, which is to choose according to analogy: there is, in fact, no way to find a unique name to designate the pounce of the squid, fish spine, and bone; there are, however, properties that are attributed to these various things as if they shared the same nature *(physis).*   *(98a20)*

We have here a relation of analogy that is, so to speak, involved in a process of "taxonomic devaluation." In fact, the pounce of the squid, fish spine, and bone are first said to be analogues in an implicit reference to the thesis which says that fish spine is to fish what bone is to bony animal, and so on (see *PA* 2.8.654a20). Then squid bone, fish bone, and bone are grouped within the same *genos,* which, although ordinary language does not give it a name, can be called the *genos* of "parts that provide rigidity to the body," as we have seen above. For, as Aristotle says at the start (98a3), he is concerned in this chapter with the construction of *gene;* and noting Plato's terminological usage in the *Sophist* and *Statesman,* he underscores the "natural" coherence of the *genos* of rigid parts by using the word *physis.*

If the biological corpus is relatively novel in its use of analogy in comparison with the logico-metaphysical books, that is not because of a conceptual difference — since analogy obeys the same rules everywhere — but because of the importance of analogy in Aristotle's biology. We must realize, then, that the use in the biological corpus of the terms "analogy" and "analogue" in the exact sense that I have just presented is not limited to anatomical and other characteristics that might be used for the construction, eventually, of a system of classification. Thus, in the *History of Animals* (8.1.588a25), Aristotle explains that psychological faculties differ between man and various other animals either by the more and the less or analogically. And this example helps us to understand the truly fundamental function of the analogical relationship in biology. It does not serve so much to set apart natural families of living things as to relate one group of animals to another by some point of reference, and ultimately to relate all living things to one unique being, taken as a model of intelligibility, man. Thus, in a passage in the *History of Animals* (1.6.491a19), Aristotle justifies the order of presentation adopted in this treatise with the following remark:

> To begin with, we must take into consideration the parts of Man. For, just as each nation is wont to reckon by that monetary standard with which it is most familiar, so must we do in other matters. And, of course, man is the animal with which we are all of us the most familiar.   (*Trans. Thompson*)

Aristotle's concern here is with what is "better known by us," according to the distinction that he uses, notably in the *Posterior Analytics* (1.2.71b33),[38] between that which is "prior and better known by us" and that which is "by nature." Thus, in a sense, all the "anatomical" parts of the *History of Animals* — most of the first four books, and to a lesser degree the other parts of the work — really do not constitute, despite what some have mistakenly said, a comparative anatomy, but are rather an anthropocentric anatomy and ethology, and that thanks to the employment, explicit or not, of analogy.

It seems to me that LeBlond proposes a one-sided interpretation when he writes:

> This originality [of Aristotle] resides rather in the perpetual application of the comparison between species and between genera, and even between kingdoms. That comparison is based on an explicit doctrine of *analogy*, perhaps of structure, but especially of function, and that makes Aristotle the initiating genius of comparative anatomy and of comparative physiology.[39]

In fact, the idea of a structural and functional unity of all living things is indissolubly tied, in Aristotle, to a hierarchical conception that makes him rank living things into an order of growing perfection. Thus, in the *History of Animals* (8.1.588b21), he writes:

Such animals are always prior by a small difference to *(pro)* the others and show already that they have more life and movement.

Man is thus the ultimate stage of this scale of beings, and therefore also the condition of its intelligibility: he is prior to the other animals as actuality is prior to potentiality. This is what explains the fact that one finds in other animals human capacities in a "germinal" state:

> The traces of these differentiated characteristics may be found in all animals, so to speak, but they are especially visible where character is the more developed, and most of all in man. The fact is, the nature of man is the most rounded off and complete, and consequently in man the qualities or capacities referred to above are found in their perfection.   (HA *9.1.608b4; cf. 8.1.588a19, 33*)

There is thus indeed a hierarchical linear disposition, neatly marked by the preposition *pro,* which we must here interpret almost spatially. Animals are, as it were, sketches of the human animal; we shall see below that Aristotle declares that, in relation to man, they are like dwarves, that is, badly proportioned and badly constructed men. It is a strange comparative anatomy that, in comparing two living forms, regards one as the caricature of the other.[40] On the contrary, it is by replacing hierarchy with true comparison that comparative anatomy was able to come into existence. Darwin's religious adversaries often blamed him for having destroyed the idea of a human realm separate from the animal kingdom, but that destruction was already a premise, even when not enunciated, of comparative anatomy.[41]

Nevertheless, as is often the case in Aristotle, especially in the biological realm, the metaphysical point of view (here, the anthropocentric doctrine of the differential perfection of living things), although it continues to be determining, is, as it were, overcome by observation. Thus, Aristotle notes the approximate character of this zoology that looks at the human as its goal: human traits still not developed are found, "so to speak," in all the animals. Aristotle recognizes, in the first place, that some living things are so distant from man that it would be illusory to look for real correspondences between the organs and functions of those entities and those of man. Second, preferring observation to finalistic speculation, Aristotle notes that on some points man is outclassed by other animals, as the *History of Animals* remarks (1.15.494b17) on the senses other than touch. Finally, there is in Aristotle a theory of the adaptation of living things to their environment, which in fact falsifies the comparison between man and the other animals, since nature does not demand the same performances from all, thus making them incomparable.

What, in the end, can we think of LeBlond's claim (1945, 41), admittedly prudently qualified, that he can discern in Aristotle a sort of transformist tendency? Rather than discussing the content of this assertion by comparing texts with each other, we should observe that this is a resurrection of the same old idea that Aristotle had taxonomic intentions. For the transformist (and the necessary comparison here is between Aristotle and Lamarck) assumes that there must be an ordering of "natural" animal families before one can look for the possible passages between these families in the course of time. In any case, LeBlond thinks that this idea of the continuity of living beings and Aristotle's supposed taxonomic project are two aspects of the same doctrine.

To complete the examination of the concept of analogy in this context, let us look at a passage that is all the more interesting because it may include an involuntary lapse in the use of the terminology under consideration.

> Things that are one in number are also one in *eidos*, while things that are one in *eidos* are not all one in number; things one in *eidos* are all one in *genos*, while things one in *genos* are not all one in *eidos* except by analogy; while things that are one by analogy are not all one in *genos*. (Metaph. *Delta* 6.1016b36)

Interpreters have read this passage too quickly, seeing in it a chain of progressively larger classes: one in number, one in species, one in genus, one analogically. In fact, the passage says that even a relationship at the level of the *eide* of the same *genos* can be said to be analogical. But this statement is terminologically un-Aristotelian, since the Aristotelian relation of analogy is between *gene;* that is, analogy supervenes on *gene.* Doubtless, that is why Ross says that Aristotle refers here to analogy "by mere inadvertence" (1924, vol. 1, p. 305). He may be right, but one might also see in this passage one of those abbreviations of expression for which Aristotle is famous. If the *eide* of a *genos* are taken in turn as *gene,* then they can have between them analogical relations. Thus (and perhaps Alexander's commentary has this right), man and horse are *eide* of the *genos* "animal," but there exist between them, if one takes them as *gene,* several analogical relationships that make them, in a sense, one. For example, one may take the nail in one as the analogue of the hoof in the other, and so on.

The first fact that strikes the reader of Aristotle's biological writings is that the philosopher almost always uses the word *genos* when he speaks of a group of animals. Balme counts only thirteen cases in which a group of animals is designated by the term *eidos.* We saw the reason for this massive use of the term *genos* for designating animal families when we examined the logical functioning of the concept: when Aristotle applies the word *genos* to a group of animals (or to any group in any domain), it is because he wants to pick it out as an autonomous group in opposition to others. Once again, that does not mean that *genos* designates a determinate degree of generality on a taxonomic scale: the *genos* of living things is opposed to that of non-living, the *genos* of birds to that of fish, the *genos* of thrush to that of falcons.[42] All have their own coherence. In designating an animal family with the word *genos,* Aristotle thus says nothing about its degree of generality. This practice inevitably disorients us post-Linnaeans.

But a *genos* is by definition divisible. When Aristotle speaks of a *genos,* he necessarily implies that it has *eide.* In the biological corpus, we find the term *eidos* where it was placed in the logico-metaphysical works: the *eidos* is a subset of a *genos*, whether expressed or understood. This logical dependence of *eidos* on *genos* is doubtless one reason why Aristotle does not use the term *eidos* for designating an animal class as autonomous. This dependence also implies that, in relation to its genos, the *eidos* is indivisible; it does not become divisible unless, by a change in levels, it is taken in turn as a *genos*. As a subset of *genos*, the word *eidos* inherits its variability: it thus cannot designate a fixed level of generality any more than *genos* can.

Many passages support this interpretation. I have chosen three of them, for two reasons that may at first seem conflicting. The first reason is that the passages are absolutely senseless if one tries to conserve the taxonomic point of view. The second is that I think the modern reader cannot rid them of all obscurity, and they thus give a good idea of the complexity of the conceptual and terminological functioning of Aristotle's biology. Thus, my interpretation aims to make some progress in comparison with its predecessors, but not to resolve all problems. The three passages are all taken from the *History of Animals.* Here is the first:

> We must now go on to describe the arrangement of the internal parts, and first of all those of the blooded animals, because this is the feature in which the greatest *gene* differ from the rest of the animals: they are blooded, whereas the others are bloodless. The former [greatest *genos*] includes man, those of the quadrupeds that are viviparous, as well as those of the quadrupeds that are oviparous: birds, fish, cetaceans, and any other anonymous group there may be; [anonymous] because the *eidos* is not a *genos* but an *[eidos]* that is simple in relation to the individuals — for example, the serpent and the crocodile. *(2.15.505b25)*

In this passage, the use of the schema *genos/eidos* is applicable on several levels. To start with, Aristotle names several *eide* that fall within the overriding "greatest *genos*" of blooded animals: man, viviparous quadrupeds, and so on. He then turns to "any other anonymous group there may be," such as the serpent and the crocodile. This text has hitherto proved to be a "road of doubt and hopelessness" for Aristotle's commentators. Let us examine one by one the difficulties it presents.

At least two difficulties arise at once. First, how can Aristotle treat crocodiles and serpents as two *eide* that are "simple" and thus "anonymous"? "Anonymous" creates no special difficulty: Aristotle does not mean that the animals in question have no name, but that one cannot group them under a family that has no name of its own. But what of "simple"? In general, Aristotle distinguishes two sorts of crocodiles.[43] One might conceivably claim that these are two varieties of a single species, which is why the group of crocodiles can properly be designated as "simple." But what about the serpents? Aristotle obviously knows of several kinds, each of which has its own name. Elsewhere he writes explicitly of the *genos* of crocodiles (*HA* 9.1.609a1) and the *genos* of serpents (*HA* 2.17.508a8).

The second difficulty concerns the "oviparous quadrupeds." If we assemble the various passages in the biological corpus that mention them, we arrive at the following list: the various sorts of crocodile, the various sorts of turtle (i.e., the land or sea *chelone* and the freshwater *hemys*), lizards, frogs and related animals, and chameleons. Scholars have been so struck by the fact that Aristotle here separates the crocodile from oviparous quadrupeds that they all have attempted to emend the text. Balme (1962, 92ff.) gives an adequate summary of those emendations before proposing his own.

Here we touch on the essential point of divergence between my interpretation and those that have preceded it. If one thinks, as previous commentators have, that this passage contains the sketch of an animal taxonomy, then it actually *excludes* the crocodile from the oviparous quadrupeds and by that very fact is in contradiction with the rest of the biological corpus. The best solution, then, is to suppose that the text is corrupt. I would suggest that most of the difficulties that have hampered previous commentators can be solved if one takes proper account of the point that the classification of animals that is put forward here makes no claim to validity in its own right. In any case, it is difficult to see what advantage Aristotle would have gained from presenting such a brief systematic in this place. Worse still, can we seriously imagine that Aristotle has here simply "forgotten" that the crocodile is an oviparous quadruped?

It is first necessary to put this passage back into its context. The passage is programmatic and a turning point: having studied the external parts of animals, Aristotle says that he is now going to emphasize their internal organs. And until the first chapter of Book 3, he will examine the anhomoiomerous parts of blooded animals, before taking an interest in their homoiomerous parts in the rest of Book 3. Then, in Book 4, he treats the bloodless animals. One must therefore read this passage, and especially the enumeration of the classes of animals it contains, in the light of an approach to the internal parts. Concerning the use of the terms *genos* and *eidos,* the only relevant question we can ask of this passage is what they signify in the context of a study of internal organs. But, *in this context,* Aristotle calls *genos* a group of animals within which the arrangements of these organs in the various subgroups admit of differences. Each *genos* is indeed divided into *eide* according to what I have called an axis of division, here that of the disposition of the internal organs. Crocodiles and serpents, no matter what differences they otherwise present in their various species and varieties (differences of size, mode of life, and so on), are identical to each other with respect to their internal organs; not absolutely identical, but sufficiently so that we may take it that they do not admit specific difference, and thus do not constitute *gene.*

This approach to understanding the texts can be taken further if we compare what Aristotle writes here of serpents with what he writes about them in another passage of the *History of Animals:*

> The *genos* of serpents differs when compared either with the above-mentioned creatures or with each other. For while the other *gene* of serpents are oviparous, the viper alone is viviparous. *(3.1.511a14)*

The first use of *genos* truly contradicts the passage 505b31, which states that serpents form an *eidos* and not a *genos,* because these two texts stem from the same perspective: that of the study of the internal parts. But a new fact has intervened: whereas for the other internal parts serpents are "*indifferent*," that is, do not present specific differences, their difference in respect to the mode of reproduction has organic repercussions that cannot be ignored, for the viviparous — or rather ovoviviparous — viper has a uterus that is different from that of the other serpents (511a17). The fineness of the division of animals is thus variable and proportional to the fineness of the anatomical study. The second use of the word *genos* — in its plural form *gene* — is, if I may say so, "banal." Aristotle is no longer dividing according to the perspective of the internal organs, but again takes up the term *genos* to designate, as usual, an entire animal family, here the various sorts of serpents.

Thus, our earlier text (*HA* 2.15), insofar as the internal organs are concerned, turns on a double division. Aristotle takes the "blooded animals" as a *genos* and distinguishes therein two sorts of *eidos*. *Eide* of level 1 include, for instance, viviparous and oviparous quadrupeds. These animal classes act as *eide* and not as *gene;* still less do they act as "greatest *gene*," as Balme seems to suppose (1962a, 93). But there are also *eide* of level 2, which Aristotle here calls "simple *eide*," such as the crocodile. We could indeed properly go so far as to say that the *eide* of level 2 are such only because Aristotle takes for granted that the *eide* of level 1 have here been taken as *gene*.[44]

This passage does not therefore in any sense exclude the crocodiles from the class of oviparous quadrupeds. All Aristotle means to state here is that the study of the internal parts can properly be carried out in relation to classes of animals, each of which has a quite different extension. Thus, we can study the disposition of the internal parts in the oviparous quadrupeds; this in no way inhibits a corresponding study of the internal parts in the crocodile. Thus, Aristotle could just as well have given birds as an example of an *eidos* of level 1, as in fact he does do, and eagles as an example of an *eidos* of level 2, without at all meaning to imply thereby that eagles are not birds. The point is that the naturalist may study birds as well as eagles, oviparous quadrupeds as well as crocodiles. This is the meaning that should properly attach to the qualification that we have translated as "and any other . . . group there may be." Obviously, Aristotle is not here wondering whether or not there exists an *eidos* of crocodiles or of serpents. By the words we have quoted, he means no more and no less than that we may, *if we wish* — that is, depending upon the degree of fineness and of exactitude that we wish to give to the inquiry we have in hand at any particular moment — concentrate our attention only on what I have called the *eide* of level 2, in order to take account of the internal parts of blooded animals.

Nevertheless, one problematic term remains in the passage that we are examining: the word *anthropos* (line 28). Aristotle seems to consider man an *eidos* of level 1, even though man would more truly than serpents, from the perspective of the internal parts, deserve the title "simple" *eidos*. This usage cannot be understood without a recognition of the great structural complexity of this passage, in which three different distinctions overlap. The first is between the groups of animals that, from the perspective of the internal organs, can be called *genos* or *eidos*. The second, on which, as we have seen, the very plan of the *History of Animals* rests, separates man from the other animals to make of him the intelligible model of the living world — so much so that the study starts from man to proceed to viviparous quadrupeds, oviparous quadrupeds, and so on, descending toward animals morphologically farthest from man: this is indeed the descending list that we find in our text. Probably the expository plan of the *History of Animals* has contaminated the list of *gene* constructed from the perspective of the internal organs, locating man there to our surprise. Moreover, a displacement occurs, whereby Aristotle arrives at a new distinction — not, it is true, in the passage that we have quoted, but in the lines that immediately follow:

All viviparous quadrupeds, then, are furnished with an esophagus and a windpipe, situated as in man; the same statement is applicable to oviparous quadrupeds and to birds, except that the latter differ according to the *eide* of [each of] these parts.   *(505b32)*

Here there is no doubt that the pair *genos/eidos* is no longer applied as before to different families of animals, but to organs: each organ of a certain type (i.e., having a determinate function) is taken as a *genos,* within which are deployed the various *eide* according to the animals considered. There is thus a "*genos* windpipe," which is divided into *eide* according to each of its properties (position, size, and so on). I shall soon try to show that in Aristotelian biology the cardinal use of division by means of the pair *genos/eidos* is that which divides organs and functions and not that which divides animals. This division of the same organ, taken as a *genos,* into *eide* is stated in the last part of the passage just cited. A. L. Peck translates thus: "except that they differ in the conformation of these parts"; and Thompson thus: "only that the latter present diversities in the shapes of these organs."

The second passage from the *History of Animals* on which I would like to rely has already been noted above. After having listed several "greatest *gene*"—birds, fish, testaceans, crustaceans, molluscs, insects—Aristotle writes:

> For none of the other animals there are great families *(gene megala),* for one *eidos* does not include several *eide;* but sometimes the *eidos* is simple and does not include a specific difference—for example, man; sometimes the groups have in them the specific difference, but the *eide* are anonymous.   *(1.6.490b15)*

The whole of this very difficult passage—490b7 to 491a6— has been made much less obscure by the detailed study by Alan Gotthelf (forthcoming). In particular, Gotthelf thinks— contrary to other interpreters—that the "anonymous" *eide* (which, I repeat, are not by that fact unnamed) are the containing and not the contained classes. One problem with this interpretation is that Aristotle would then call a class an *eidos* because it is divisible, which he does nowhere else. I recognize that a decisive answer to this question is probably impossible. It does seem possible, however, to make sense of the statement "one *eidos* does not include several *eide.*" Gotthelf, while recognizing that this part of the sentence relates to the proof that the term *eidos* can be used at several levels of generality, ultimately concludes that there is in this passage a "popular," or nontechnical, usage of *eidos.* I prefer to see in this statement a negative definition of *eidos:* in a given context, an *eidos* is not divided into *eide.* Notice that here man is called an *eidos.* We are sufficiently far from the taxonomic reading not to be surprised by this lack of harmony with the preceding passage. But as soon as Aristotle has said that, from his present viewpoint, man is an *eidos,* he feels the need to assign this *eidos* to a *genos* that includes it:

> The *genos* of animals that are simultaneously quadruped and viviparous has many *eide,* but they are anonymous. Each of them, so to speak, is called by the name of an individual, as: man, lion, elephant, horse, dog, and so on. *(490b31)*

Notice also that in this passage Aristotle ranks man among the viviparous quadrupeds, while in other passages he distinguishes him from them. That proves — and obviously it is fundamental for my study — to what degree Aristotle constructs orderings of animal families for the immediate occasion. For on the one side, man is the limiting case of quadruped, or (and this manner of putting it would be more correct from an Aristotelian point of view) viviparous quadrupeds are degraded copies of man — so much so that "the viviparous quadrupeds have front feet, analogous to arms" (*HA* 2.1.497b19). But on the other side, when the study becomes more precise, Aristotle emphasizes the differences between man and the quadrupeds. Hence the numerous passages in which man is not presented as an *eidos* of the *genos* of viviparous quadrupeds. Thus, the *Progression of Animals* tells us that, in respect to the flexing of limbs, man is "contrary" to the viviparous quadrupeds (1.704a23; cf. *HA* 2.1.498a3).

The third passage to which I would like to refer is the following:

> The *genos* of birds as a whole drinks little, and crooked-taloned birds
> never drink at all, except a small *genos* and they rarely; that is particu-
> larly the case with the kestrel. The kite has been seen to drink, but it
> drinks rarely.   *(8.3.593b28)*

At first sight, it would seem that here the *genos* is subdivided into
gene. But in fact there is no division in this text: Aristotle does
not here distinguish several ways of drinking and not drinking,
or several *eide* of organs meant for drinking. Indeed, entirely
naturally, as is his habit, he names with the same term, *genos,* just
a few words apart, two families with such different extensions as
the class of birds in general and a subdivision of crooked-taloned
birds.

Now we are much better armed to examine the cases, gathered by Balme, in which *eidos* designates groups of animals. In
fact, the only valid question that remains, from an Aristotelian
point of view and not from a taxonomical point of view foreign
to him, is that of figuring out why Aristotle does not designate
these groups of animals by the term *genos.* Even so, the fact that
these groups are situated at different taxonomic levels, far from
embarrassing us, only confirms what we have already established: *eidos* does not designate a fixed level of generality. In
these instances, I shall distinguish six different uses.

1. In the first place, there are four texts in which *eidos* is given
   purely and simply as a subset of a *genos:*

   · *HA* 1.1.486a24: the *genos* fish or the *genos* bird has many
     eide.

   · *PA* 4.5.679b15: there are many *gene* and *eide* of testacea.

   · *PA* 4.8.683b26, 28: each of the four *gene* of crustaceans has
     several *eide.*

   · *GA* 3.9.758b9: the caterpillars are an *eidos* (of the *genos*)
     larvae.

2. In the *Generation of Animals* (1.11.719a7), Aristotle says that the ovovipara simultaneously partake of two *eide:* the vivipara and ovipara. This is a remarkable illustration of the doctrine of the *genos* as a unity of contraries: here, too, the *genos* considered is not a group of animals but a biological function, reproduction.

3. Several texts call groups of animals *eide* because they are presented as results of one or more divisions, even though that may not always be immediately obvious:

· *HA* 4.7.532b14 notes that there are many *eide* of cicadas and indicates two axes of division (i.e., two specific differences) according to which these *eide* can be isolated: largeness/smallness and the faculty of singing/silence (cf. 4.9.535b8 and 5.30.556a14).

· *HA* 9.24.617b16: there are three *eide* of daws (crows), and Balme is surprised that, two lines later, Aristotle speaks of a *genos* of Lydian and Phrygian crows; but there is, here too, a double division: crows are divided into Lydian crows and crows elsewhere — those familiar to us — and these are divided into several *eide.* Two lines below that, however, Aristotle writes that there are two *gene* of "korydales" (larks?): those that run on the ground, have a crest, and live in an isolated way, and those that are gregarious. We should note that, in choosing these characteristics, Aristotle does not use the diairetic process that divides the *genos* into *eide* according to contraries; not opposing their characters two by two (according to the axes of division), Aristotle here treats the families of "korydales" as relatively autonomous entities, and thus calls them *gene* rather than *eide.*

· *HA* 8.3.592b18: there are several *eide* of tits; but Aristotle has just distinguished birds that eat meat and worms, and in this group there are several subgroups, including the tits, so there is a double and even triple division.

The last three examples tend to prove that when he has made a division of a *genos* into *eide,* and thus finds himself at the level of the *eidos,* Aristotle can continue to call *eide* the subgroups obtained by division of the first *eide.* In fact, to return to the last example noted, two *eide* of tits are distinguished by the location of their habitation, and they (being of the same size) are opposed to a third *eidos* of tit that is smaller in size.

4. A passage in the *Parts of Animals* (4.5.680a15) would be particularly distressing if one had adopted the point of view of the taxonomists. Aristotle writes, in fact, that there are several *gene* of sea urchins, and he "proves" that assertion by the following remark, which editors have bracketed: "for there is not one unique *eidos* of all the sea urchins."[45] It seems to me that there are three steps in this passage, of which the second is to be understood as follows:

(a) There are several kinds of sea urchins (as usual, Aristotle designates them by the term *genos*);

(b) if one considers the set of all sea urchins as a *genos* (this proposition is understood because it is necessarily required by the following),

(c) then one would perceive that there is no single *eidos* of sea urchin, i.e., that the specific difference does indeed play a role within the *genos* of sea urchins, which thus fully deserves its name of *genos.*

Actually, the last proposition proves the first, according to the principle "whatever does more can also do less," for if not every difference is a specific difference (according to the contraries), if there *is* a specific difference (i.e., a division of a *genos* into *eide*), *a fortiori* there is a difference.

5. A passage in the *History of Animals* (5.31.557a24) resembles the preceding example: Aristotle notes that there is a unique *eidos* of sea lice, while a few lines previously (a4) he speaks of the *genos* of lice called "wild." We shall see that this use of *eidos* confirms the validity of the logical schema

of division of the *genos* down to one of its limiting forms. In fact, the preceding passage posits a division into air lice and water lice. Thus, there is a division according to a specific difference, which yields two *eide*. So when Aristotle goes on to divide these latter groups, we can expect one of two equally possible moves:

(a) Either Aristotle takes these *eide* as *gene* and redivides them according to the specific difference, thus obtaining *eide* of rank 2: this is what he elsewhere calls, as we have seen, "taking the difference of the difference";

(b) or Aristotle subdivides these two groups (air lice and water lice), but simply by distinguishing them by some characteristics without bringing into play the specific difference; and we have seen above in our third rubric that these subdivisions could keep the name *eidos*.

I believe that Aristotle chose the first alternative: he indicates that he has run up against a limiting case, since the *genos* of "marine lice" does not have several *eide*, but only one. That is shown by the very structure of the passage: "But there is only one *eidos* of lice in the sea." *Eidos* here leaves open the possibility of a second division even if it cannot take place: far from being careless about his vocabulary, Aristotle rather seems to me to insist upon a rigorous application of the diairetic schema until it reaches its ultimate limit. If one reads this passage according to the second alternative, in fact, it leads back to a case of the third rubric.

As for the *genos* of lice called "wild," that is examined before the division air/water is posited, and it is a *genos* from the fact of its own coherence.

6. The two remaining passages do not constitute major problems for anyone who has accepted the argument for the absence of a fixed taxonomic function for the terms *genos* and *eidos*. Nevertheless, we must recognize that we cannot be sure of their precise interpretation.

· *HA* 8.3.592b7: there are two *eide* of vultures, small-white and large – ashen-colored. Nevertheless, at 592b1, Aristotle notes that there are two *gene* of eagles.

· *HA* 9.20.617a18 refers to three *eide* of thrushes, while a few lines earlier Aristotle says that there are two *gene* of blackbirds (617a11) and three *gene* of herons (617a2).

In these last two cases, the *eidos* is not presented as the result of a division, so these two passages cannot do anything to confirm the interpretation of the taxonomists. Perhaps, in the first instance, if we consider the other passages that speak of various *gene* of eagles (notably *HA* 9.32), we might understand that Aristotle wants to suggest that the distance is less great between the various sorts of vultures than between the various sorts of eagles. In the second case, perhaps the use of the word *eidos* is a literary device meant to avoid excessive repetition of the term *genos*. However that may be, the reason for the use of *eidos* in these two cases ultimately remains obscured by the fact that, where there is no division, Aristotle has no fundamental reason not to use the term *eidos,* since there is nothing theoretically opposing its use.

Summarizing our results concerning the use of the terms *genos* and *eidos* as applied to animal families, we may say that the normal term for designating these families is *genos,* and that Aristotle, in the rare instances where he uses *eidos,* generally wants to indicate by this that there has been a division of a *genos.* But neither of the two terms, *genos* or *eidos,* indicates a constant degree of generality on which a taxonomic construction could be based. The Aristotelian doctrines of difference "according to the more and the less" and of analogical difference cannot furnish a fixed point, any more than can the notion of "greatest *gene.*" The logical subordination of *eidos* to *genos* that one finds in the logico-metaphysical writings is respected in the biological corpus. But is that enough to establish the conceptual unity hypothesized at the beginning of this chapter? It seems, on the contrary, that in being applied to living things the pair *genos/ eidos* has "lost" one of its principal logical properties, namely, the contrariety of *eide.* How can we reconcile the thesis, often affirmed by Aristotle, that horse, man, and dog are *eide* of the *genos* "animal" with this contrariety of *eide?* Are there, then, *two* doctrines of the division of the *genos* into *eide,* one "logical," the other "biological"? To show that this thesis of two doctrines cannot be sustained, it will be enough to refer to two passages of the *Metaphysics.* In Book Iota (8.1058a3), Aristotle explicitly gives man and horse as examples of *eide* of the *genos* "animal," while in that same Book Iota, both before and after this passage, he shows in the greatest detail that within the *genos* the "specific difference" concerns contrary *eide.* Also, at Delta 6.1016a25, Aristotle notes that the *genos* is divided according to contrary differences, and then, in the following line, he gives as an example of that thesis the *genos* "animal" divided into "horse, man, dog."

This problem will force us to abandon, once again, a position that seems to us so "natural" that we do not even think of doubting it. In fact, until this point, I have, like all other commentators, generally taken it for granted that in Aristotle's biology the terms *genos* and *eidos* are primarily applied to animal families. But that is doubly false. In the first place, it is false quantitatively: even if the word *genos* is normally used to designate an animal family, no matter what its degree of generality, the *genos/eidos* conceptual scheme is rarely applied to animal families. In the second place, I propose to show in the next chapter that it is also false qualitatively — namely, that the Aristotelian diairetic process is not applied fundamentally to animal families, but to the "parts" of animals. When he uses it apropos of animal families, Aristotle does so in a derivative and weakened way. Once again, it is *we* who attribute theoretical priority to the division of animal families. This is yet another trace of the taxonomic prejudice, so difficult to extirpate.

The fact that Aristotle usually uses the word *genos* to designate animal families raises a problem: namely, that, in Aristotle's writings, the word *eidos* simultaneously designates (1) a subset of *genos* and (2) "form" in opposition to matter — that which makes a being what it is. But it is indeed this form that is conserved in the reproduction of living things: "Man begets man." It would seem that the traditional interpretation that translated *eidos* sometimes as "species" and sometimes as "form" was coherent, and that would again promote Aristotle to the rank of precursor, for is this intelligible structure that is the form not a prefiguration of the notion of genetically transmissible inheritance?

In fact, Aristotle uses the word *genos* to designate the location where organic form is conserved. This shows us again, and in a different way, how far he was from the concepts of modern systematics, and especially the notion of a species.[46] In fact, the essential feature of the concept of species for modern biologists is the conservation and transmission of a morphological, physiological, and behavioral structure. In using the word *genos* to designate, among other things, our species, Aristotle seems, on the contrary, to insist on change rather than conservation. Recall here my analysis of the *genos* as the place where contraries are deployed within the Same, as well as the derivation of *genos* from *genesis.* The eidetic identity of every living thing is extended in time between the beginning of the actualization of the potential *eidos,* which the sperm carries from the father, and the destruction of that *eidos* by death. The male parent, in fact, carries in his seed a potential *psyche* (e.g., *GA* 1.1.735a8; *PA* 1.1.641a23), and we know that "the *psyche* is *ousia* in the sense of form *(eidos)* of a natural body" *(De An.* 2.1.412a19). In several places, Aristotle also reminds us that between the living thing and its cadaver there is only a homonymous identity (e.g., *Mete.* 4.12.389b31). But this entrance into and exit from life are truly *genesis,* movement according to *ousia,* because such movement is effected from one contrary to another. And between the two extremes of fertilization and death, one finds, other than the numerous risks of errors, a struggle between the form and the matter that is far from producing a uniform movement of growth. Without returning to Aristotle's teratology, I could give as evidence Aristotle's observations on the growth of children and his speculations about their resemblance to their parents.

Aristotle was struck by the fact that not all the parts of children's bodies grow continuously and at the same rate. Thus, he says several times that children are like dwarves, with the upper part of the body more developed than the lower part.[47] He also says, in the *Parts of Animals* (4.10.686b3), that "all the other animals are dwarves by comparison with man." We find here once again the Aristotelian notion of man as the completed element in the animal series. This equivalence of children and other animals as opposed to adult human beings has the consequence that the individual human being recapitulates during his lifetime differences in the *genos* of living things. Perhaps it was this sort of idea that earned for Aristotle the approbation of Haeckel, noted in the Introduction of this study.

The problem of the resemblance of children to their parents interests me even more, since what is at stake is knowing if the *eidos* of the father will be transmitted as such, or in what measure it will be altered. Aristotle presents several cases, describing a kind of descent toward monstrosity, the steps of which, according to the *Generation of Animals* 4.3, are as follows:

1. The child resembles its father.
2. The child resembles its mother.
3. It resembles its ancestors.
4. It resembles no one, though it has human form.
5. It does not have a human appearance and is a monster.

Each of these steps breaks down into possible subdivisions according to whether the resemblance is found "for the body as a whole or for each of its parts" (767b1). According to the degree of concoction of the female menstrual residue (which is the matter for the offspring) and the force of the movement that the masculine sperm is able to impress on this residue, the product of generation will find itself at a higher or lower level on this scale of similarities: "The first separation is that the offspring is a female and not a male" (767b8). While, on the one hand, Aristotle's constant doctrine is that in generation the father carries the *eidos* and the mother the material, nevertheless, on the other hand, he writes here that these successive deviations that lead from the same (the son resembling his father completely) to the other (the monster — that which does not even have the human *morphe*) involve a progressive separation from the *genos:* "In [monsters] nature has in a way separated from the *genos*" (767b7). Nowadays we are more sensitive than Aristotle to the identity of what is transmitted from one generation to the next, so that we call "human" every offspring of a pairing of human beings. For Aristotle, in contrast, the plasticity of the human *genos* (as of every other animal *genos*) has limits, and certain monsters must doubtless be placed beyond those limits. Although we have no text that says so explicitly, the coherence of Aristotle's interpretation of formless monsters allows us to take them as heterogeneous, from a human point of view, for the *genos anthropos* is unable to accept their excessive difference.

As for offspring whose humanity is indubitable but who do not resemble their ancestors, but "any chance person" (767b2), they are, one may say, "generic" infants, in that "there remains in them only what is common to all men" (768b11). In a sense, they are more the realization of the universal (768b13) than of a particular (b15).

Aristotle thus conveys by the term *genos* the transmissible type that in our eyes characterizes the species, and by *eidos* the model that is actually transmitted in generation. It would be necessary for these two notions to converge and become superimposed for the modern concept of a species to be born. For Aristotle, the species did not yet exist.

What, then, of the problem posed at the beginning of this chapter: whether Aristotle's logic came from his biology, or the reverse? I cannot resolve that problem simply on the basis of what I have shown in this chapter, but I can at least reject the interpretation of those who, like LeBlond, think that Aristotle was a logician of concepts because he was a systematician of living things. LeBlond's idea is nevertheless attractive for two reasons. The first reason, quite general, is the tendency, so frequent among historians of philosophy, to underline the originality of the author whom they study — and often, more importantly, whom they love — in opposing him to what preceded. What is more fascinating than a great innovator? Thus is born the astonishing legend of an Aristotle who was simultaneously biologist and vitalist, turned toward the concrete in opposition to the mathematical Plato who sought his models of intelligibility in celestial mechanics. Even more astonishing is the vitality of such a legend. Not that this notion is false: it is beyond true and false, in this purely schematic form. One could as well point out that Plato animated the cosmos far more than Aristotle ever did and that the cosmic vitalism that one sometimes finds in the *De Caelo* is certainly of Platonic origin. As for the epistemological vitalism (some say "biologism") of Aristotle, it arises from undue haste on the part of the interpreters: because Aristotle was very interested in biology, it has been entirely natural for commentators to think that he was looking for models of intelligibility in vital processes. But if we look at the texts, we notice that "technomorphic" models of explanation are much more numerous than biological analogies: nature certainly wins out over art, but it is nevertheless art that permits man, the technical animal, to comprehend nature. Finally, without despising the merits of Plato, notably in the *Theaetetus,* it is high time to recognize Aristotle's eminent place in the history of mathematics, especially after the "crisis of the irrationals." In the words of Jean-Toussaint Desanti (1967, 440):

> [This crisis] not only extended the domain accessible to calculus. It demanded a reconstruction of the entire mathematical edifice and a rebuilding of the notion of a mathematical entity. It broke the traditional frame (Pythagoreanism and Platonism) in which, in part, because of the limitations of their mathematical technique, . . . the Greeks at first tried to conceive number; it called for another

philosophy of mathematics whose elements were put together by Aristotle.

Another cause of LeBlond's interpretation is doubtless the reaction of an historian of philosophy to the seduction of a triumphant contemporary philosophy — in his case, Bergsonism. To show that people in style were wrong to claim discoveries, and that Aristotle had anticipated them, would not be disagreeable to the Reverend Father LeBlond. In this same search for "precursors," Gilson "showed" that Thomas Aquinas was the first existentialist.

In keeping Aristotle out of today's disputes, I am trying to give his thought a less one-sided image. For the question that occupies us, which springs most clearly from the preceding analyses, is the tight *solidarity* of his logical and biological studies — a reflection, doubtless, of their contemporary character during Aristotle's life. We must continue in the strict desire to import into Aristotle as few as possible of our own questions as we consider his zoological classifications. For Aristotle, although not a taxonomist, did not fail to classify animals.

3

THE STATUS AND FUNCTIONS OF ARISTOTLE'S ZOOLOGICAL CLASSIFICATIONS

THE COMMENTATORS DID NOT invent Aristotle's desire to classify animals: the biological corpus includes many passages in which he divides living things and distributes them into distinct groups. What the commentators have added is that Aristotle's classifications were aiming at that modern "natural" classification called "taxonomy." Now I must definitively establish that no sketch of taxonomy can be found in Aristotle's classificatory passages. I shall go on to try to see in these texts their perspective, in both senses of that word: both their goal and the scope of the theoretical viewpoint that they outline. Thus, I shall also review the logical structure of Aristotle's biology as a whole.

Here is a list of several of Aristotle's classifications concerning all or most animals, leaving to one side the many small "regional" classifications with which Aristotle's biological work is peppered:

- *HA* 1.1.487a11: "Animals differ from one another by mode of life, actions, habits, and parts." These four points of view are developed extensively throughout the *History of Animals*.

- *HA* 4.8: Aristotle proposes a sketch of a classification, or at least a distinction between animals according to the presence or absence of various sense organs.

- At the beginning of Book 5 of the *History of Animals,* Aristotle orders animals according to the increasing complexity of their reproductive organs: we must take them in the following order (539a10): testaceans, crustaceans, molluscs, insects, viviparous and oviparous fish, birds, oviparous and viviparous terrestrial animals.

- In several places (notably *HA* 8.1.588b4), Aristotle classes living things, vegetable and animal, according to the extent and complexity of their vital functions. I have already noted how this "scale of beings" has man at the top: in the *De Anima,* there is a sketch of a classification of animals according to the parts of the soul that each possesses (2.2.413b32, 3.414b33).

- Several passages outline a classification according to mode of reproduction: viviparous, oviparous, ovoviparous. But this very general criterion is sometimes combined with others: in the *History of Animals* (3.1), with the arrangement of the reproductive organs; in the *Generation of Animals* (2.1.732b15 – 733b16), with the degree of perfection of the product of generation.

- We also find in Aristotle a distribution of animals, starting from a unique central schema, according to the form and arrangement of the various organs proper to the various vital functions. This sort of passage leads to the vast anatomical and physiological comparisons of which I have already spoken: thus, in the *Parts of Animals* (4.10), Aristotle goes on to a kind of partial reconstruction of the animal kingdom on the basis of the central schema mouth-neck-stomach. We shall examine this method of *a priori* reconstruction of living things in relation to a famous passage in the *Politics* (4.4).

Perhaps the fact that Aristotle has taken so many different standpoints ought to be enough to show that he did not have a taxonomic project. But the commentators — and as I have shown in the first chapter, these include G. E. R. Lloyd — have often preferred to see in the various classifications proposed by Aristotle tentative steps toward a definitive and universal classification. That is also the position of Pierre Louis.[1]

In fact, these classificational passages are not taxonomic because they do not distribute animals into the slots of a unique and fixed construction, but are rather orderings developed for the occasion, adapting their extension and rigor to the needs of the exposition in progress. I shall present four arguments to show that these passages are not taxonomic.

First, Aristotle never explains how he has established his classificatory criteria, but rather posits his distinctions without justifying them. One may find in Aristotle, especially in certain cases of ambiguous animals, traces of the difficulties he encountered in delimiting the different animal families he names. But we never find him raising, at the theoretical level, the problem of the distinction of these families; for Aristotle that is obvious. Thus, in the *History of Animals* (4.1.523b1), he writes that there are several *gene* of bloodless animals, and he enumerates them, describing each *genos* in two or three lines: molluscs, crustaceans, testaceans, insects. He does not say in this passage whether it is an exhaustive list, but he does not question this nomenclature. However, the list appeared to Cuvier, and appears to us, to have great taxonomic value. In the *Parts of Animals,* in contrast, Aristotle declares that this list is exhaustive, but without giving the least justification (*PA* 4.5.678a30; see also *GA* 1.14.720b4). Modern interpreters have indeed speculated that Aristotle constructed his animal families inductively, but we search in vain in Aristotle's works for an account of any such inductive construction of animal families.

The second argument relies on an account of the very sort of difficulties that Aristotle intends to resolve with the assistance of the classifications he develops. Thus, several times in Book 1 of the *Parts of Animals,* he asserts that the naturalist ought to group animals in order to avoid repetitions, since what he proposes to study are the common functions of all animals, or of many among them — as, for example, sleep, respiration, and locomotion. Thus, his goal is to save time and make the exposition less tedious, but the grouping of animals into families does not in itself constitute a theoretical advance. This purely instrumental character of the classifications, which denies them all theoretical autonomy, appears particularly clearly in a passage at the beginning of Book 5 of the *History of Animals.* Alluding to the various divisions of the animal kingdom that he has proposed, Aristotle writes: "Since the *gene* have previously been divided, we ought now to try to follow the same path in our study [of the organs of generation]" (5.1.539a4). But he continues: "except that, in that case [i.e., the preceding study of the functions and organs], we began with the study of the parts of man, while now we must end with him because he is the most complex subject." This reversible classification, according to the difficulty of the enterprise (*pragmateia,* 539a8), has indeed a pedagogical and not a theoretical function. And Aristotle does not oppose, in this case, the pedagogical to the theoretical, as when he distinguishes what is better known "by us" and "in itself." He recognizes that beginning with man, the "best-known currency," had only a practical justification; he does not now say that the study that starts from the simplest organisms is done according to any "natural" order. For Aristotle, in this case, there is only the method that suits the occasion.[2]

I draw my third argument from the very order of exposition leading up to the division into *gene* that provides Aristotle with the general plan of the first four books of the *History of Animals.* From the beginning of Book 1, it is quite clear that the task of classifying animals has no priority for Aristotle. In fact, he is led to mention the differences between animals as a consequence of considering the differences between the *parts:* "Some animals resemble each other in all their parts, others differ" (486a14).

The similarity/diversity of the parts is prior, for the zoologist, to the similarity/diversity of animals. That is just what is summarized in this remarkable formula: "According to the parts that each of the animals possesses, that is how they are simultaneously other and the same" (486b22). Subsequently, touching on the problem of the division of animals, Aristotle gives the several criteria of distinctions already noted above: mode of life, activities, habits, parts (1.1.487a11). We can see clearly that these distinctions are only empirical considerations, that this division into four is simply rhetorical: it is indeed necessary to introduce some order into the mass of assembled facts to be able to present them. And here Aristotle gives us these criteria for classification telling us neither where he has gotten them nor whether the list he gives is exhaustive. This division is, furthermore, open to revision: thus, when he is working on ethology, Aristotle feels the need to subdivide the large divisions that he has established. In Book 8, for example, the activities of animals are divided according to two points of view, which could allow the construction of a table with two entries (12.596b20): (1) in relation to their goal (coupling, care of offspring, search for food), and (2) in relation to temperature and seasonal conditions (some combat cold by staying put, others emigrate). This is only a refinement of a division already posited. In some cases, in contrast, the criteria of division already established turn out to be not very useful, and this necessitates a reexamination in the course of the exposition. Thus, after having stated the division between animals that are fixed *(monima)* and those that are mobile *(metabletika)* (1.1.487b6), Aristotle remarks that there are some animals that sometimes remain fixed and sometimes move — for example, sea anemones. But this remark does not lead him to revise the distinction that proved to be only approximate. Furthermore, when new divisions appear in the course of the exposition, Aristotle simply announces them at the moment when they arise. Thus, in the same Book 8 of the *History of Animals,* in order to open a new field of study after he has ended a kind of little treatise on animal pathology, Aristotle writes

simply that "animals also *(de)* differ according to places"
(8.28.605b22).[3] The order of exposition in the *History of Animals* also leads me to observe that there is an obvious lack of
symmetry between the four rubrics used to distinguish animals,
since the parts are far more important than the other three: mode
of life, activities, and habits. Aristotle's entire biology proves
that. In fact, after taking up a few general considerations about
the other three distinctions, and after making some methodological remarks, at 491a6 Aristotle sets to work precisely on the
study of the parts. He insists on the empirical character of the
order of this study, which takes the shape of the human body as
its line of attack:

> So first of all we must consider the parts of which animals are
> composed; for it is in respect of its parts, first and foremost, that any
> animal as a whole differs from another: it may be that one animal has
> parts that another lacks, or that the parts vary in their position
> or arrangement, or according to the differences we have already
> mentioned, according to *eidos,* i.e., excess or defect, or analogy, i.e.,
> according to the contrariety of characteristics. But we must first
> consider the parts of man; just as each person counts in the currency
> he knows best, so also in other fields. But man is the animal best
> known to us.   *(1.6.491a14)*

This passage, which I have quoted twice before in part, thus
justifies in advance the inversion of the order of study from
Book 5 on. We may draw a further consequence from this third
argument: the empirical character of the distinctions posited by
Aristotle is not a second-best, a consequence of a theoretical
incapacity on his part. That is shown by the fact that he sometimes does, on the occasion of a particular study, posit the objective and theoretically justified bases of some empirical distinctions. Thus, in the passage from the beginning of Book 9 of the
History of Animals (1.608b4) in which he discusses the priority of
man in relation to other animals, Aristotle indeed furnishes a
possible approach to an anthropocentric comparative animal
psychology, and from that a criterion of an eventual classification of animals depending on whether they are closer to or

farther from the human model. In most cases, then, Aristotle leaves the distinctions that he uses in their empirical state, because for him the problem of their foundation *is not posed.*

The fourth argument, finally, by which I would like to show that Aristotle had no taxonomic goal ought to be the most convincing to those who hold that Aristotle was a systematician. It is based on problems peculiar to taxonomy — problems rediscovered *as such* by later systematicians — which Aristotle is forced to mention in his exposition because they flow necessarily from his diairetic practice, as gross and empirical as it is. The best example of this is perhaps the "overlapping of groups." This is one of the expressions that Aristotle uses to designate what I called, in Chapter 2, "groups somehow perverse," and I listed, in a note, other expressions or terms of the kind. But the term "overlapping" (*epallaxis,* sometimes replaced by the verb *epallattein*) is purely descriptive and opens no path to any theoretical reflection on the status of inter-generic classes. And, in any case, Aristotle uses this same word outside biology: in the *Politics,* for example, he uses it to signify that two concepts partially overlap (apropos of slavery [1.6.1255a13]; of acquisitiveness [1.9.1257b36]; of the fact that political constitutions have some common points [4.10.1295a9: tyranny and kingdom; 6.1.1317a2: in general]). A taxonomist obviously could not have left this problem alone after having noticed it. That does not mean that for Aristotle this "overlapping of groups" is not a *theoretical* problem: but that problem is situated at the metaphysical and cosmological level of sublunar imperfection. In biology, on the other hand, it constitutes a purely practical problem. We can see this clearly even in the very words with which Aristotle notes the ambiguous status of the seal at the end[4] of the *Parts of Animals* (4.13.697b1) — that is, in an eminently theoretical work, in which Aristotle goes beyond stating the facts to ascend to the causes:

> Similarly [i.e., like the cetaceans, which Aristotle has just said are in a way both terrestrial and aquatic, 697a29], seals and bats share in both sides *(epamfoterizein),* the first between aquatic and terres-

trial, the second winged and terrestrial; thus, they belong to both groups as well as to neither.

The Greek *kai,* which I have translated "as well as," shows that Aristotle leaves to the reader the choice of the class into which he means to rank the animals in question, according to the study that he intends to carry out.[5]

Thus, I can complete what I said in the first chapter about Aristotle's preference for commonsense divisions. I there insisted upon the polemical anti-Platonic character of that position, showing that for Aristotle "people" are closer to the truth — that is to say, to *ousia* — than the scholars of the Academy. Oddly, Aristotle remains satisfied with a rough-and-ready classification, although he occasionally had to correct the observations of practical men. But we see that for his purpose those orderings were sufficient. Sometimes Aristotle even says explicitly that his divisions are approximate; thus, in the *History of Animals,* when he needs an ordering that will be empirical but sufficient for his study of how birds nourish themselves, he writes: "All, so to speak, of the birds look for their food on land, or live around rivers and swamps, or around the sea" (8.3.593a24).

Thus, in none of these passages from the biological corpus, in which Aristotle sorts animals into distinct groups, does he unveil a taxonomic project. These passages class living things according to a multiplicity of viewpoints, sometimes anatomical, sometimes physiological or behavioral, without any of these viewpoints gaining a privileged position, since each of these classifications is proposed on the occasion of some particular investigation in progress. And when modern naturalists praise the systematic value of Aristotle's zoological distinctions, that is an anachronistic reading of these passages, because these naturalists read into them a goal that is their own. But this does not mean that Aristotle's distinctions do not possess the objective coherence that those naturalists have recognized. Like living organisms, great works adapt themselves to new environments in bearing fruit inconceivable to their own authors. One may even say that they do not acquire true descendants except at the price of a series of betrayals. In that sense, Aristotle "invented" animal systematics just as Sophocles "invented" the Oedipus complex.

There is, however, a passage that, although it is outside the biological corpus, seems to set itself the theoretical task of classifying animals. This often-cited passage is from the *Politics* (4.4.1290b25), where Aristotle tries to determine the number of possible political constitutions by examining how the different sorts of animals should be counted. Incidentally, if he should find the number of possible constitutions, he would also find the number of actual constitutions, since he thought that, in this domain, "almost everything has been discovered" (2.5.1264a3). In order to make commentary easier, I translate the passage in nine numbered phrases:

(1) It is as if we were to decide to take the sorts *(eide)* of animal;

(2) we would determine *(apodiorizomen),* in the first place, what it is necessary for every animal to have

(3) (for example, certain sensory organs, and that which digests and receives food, as the mouth and stomach, and besides the organs by which each of them moves),

(4) and if these parts are indeed the only ones, but present between them some differences *(diaforai)*

(5) (I mean, for example, that there are several sorts *(gene)* of mouth and of stomach, and also of sensory organs and of organs of locomotion),

(6) the number of their conjunctions will give necessarily a plurality of kinds of animals *(gene zoion),*

(7) (for the same animal cannot have several varieties *[diaforas]* of mouth or ears),

(8) so that when one has taken all the possible combinations, that will give the species of animal *(eide zoiou),* and there

will be as many species *(eide)* as conjunctions of the necessary parts;
(9) it is the same for the political constitutions that we have
dealt with.

D'Arcy W. Thompson (1913, 25) claims that a passage like
this, in Aristotle's time, was absolutely a novelty: "It has since
been a commonplace to compare the state, the body politic, with
an organism, but it was Aristotle who first employed the meta-
phor." Whether or not Aristotle was the first, the comparison
between biological and social organisms certainly had then a
force that usage has caused it to lose. Far from being, like Robert
Joly, "consternated" by such comparisons,[6] the historian of philosophy should learn much from them, and especially here.
From this methodological comparison between biology and
politics, he should deduce that Aristotle intends his diairetic
method to be applied to very diverse domains. That again confirms the thesis that I have argued above, that there is no reason
to distinguish a "logical" from a "biological" sense of the terms
genos and *eidos* in Aristotle. This passage also clearly confirms
another point I established in Chapter 2: namely, that *genos* and
eidos are not attached to any particular level of generality, since
Aristotle uses the word *eide* when he speaks of the subdivisions of
"animal," which is then taken as a *genos* in (1) and (8); on the
other hand, he uses the term *genos* to designate the first divisions,
whether of the kinds of mouth in (5) or of animal in (6).

The two most recent French translators of the *Politics,* Jules
Tricot and Jean Aubonnet, speak, concerning this passage, of a
"deductive method," opposed, writes Aubonnet, to "the inductive method ordinarily used in the biological treatises."[7] The
word "deductive" is rather unfortunate, for it does not correspond exactly to any Aristotelian term, and in the present case it
is only very approximately adequate. Even if the translation of
the word *epagoge* by "induction" has to be accompanied by explanations, at least there is from one language to another a
correspondence of terms. That is not the case for "deduction":

the only kind of deductive reasoning that Aristotle formalized was the syllogism, of which "demonstration" *(apodeixis)* is only a subset, for as the *Prior Analytics* says, "demonstration is a kind of syllogism, but not every syllogism is a demonstration" (1.4.25b30). The idea that Aristotle's habitual method in biology was inductive is such a common notion among commentators, who usually accept it as obvious, that we have every reason to be suspicious of it. In fact, I have already noted that an examination of the texts does not verify this notion even where it should have been the case: we find no trace in Aristotle of an inductive construction of animal families.

As our passage does not have a syllogistic form, we cannot call it "deductive," and we have to see just how Aristotle proceeds. In the first place, at (1), it would seem that he wants to divide "animal" into its *eide,* that is, by taking it as a *genos.* But that is not what he does, for if that were what he was doing, he would distinguish contrary *eide* within the *genos* under consideration, which is not the case. Instead, using the method that we may also find in the *Posterior Analytics* (2.13) and in the *Metaphysics* (Zeta 12), he constructs the definition of "animal" by stating its essential predicates.[8] That this is what Aristotle is doing is shown especially by the use of the two words "determine" *(apodiorizein)* and "necessary." The first indeed indicates a process of defining,[9] with a nuance, difficult to clarify, added by the prefix *apo:* perhaps Aristotle adds it to insist on the idea of separation that one finds in every definitional process, thus signifying that it is by abstraction that one delimits the vital functions that cannot be considered as *ousiai* (since we have seen that definition is really applied to *ousia*). As for the term "necessary," it shows that what is really at issue is exhibiting the essential attributes of the defined thing — in this instance, the functions that lead us to label an entity as animal. For it is Aristotle's constant teaching that definition, the formula of essence, is constructed on the basis of essential and thus necessary characteristics of the object defined.

One might object here that in this definitional process there is no trace of *division,* whereas we have seen that division is the primary tool for constructing definitions. But in the present passage, Aristotle does not exhibit a process of definition: he does not tell us how he has obtained the essential predicates that he lists for us — he simply says that they are essential — nor does he tell us whether the list is exhaustive, nor if these predicates are listed in the correct order. But as the *Posterior Analytics* makes clear:

> To construct a definition with the aid of divisions, one must aim at three things: to take the predicates that are in the essence, to put them in order according as one is first or second, and also to take them all.   *(2.13.97a23)*

In fact, the phrase "we would determine," in (2), alludes to a method considered well known — namely, constructing definitions with the aid of divisions. But this passage, which is written as a contrary-to-fact conditional, does not intend to show how this method is actually put into operation.

If it was not Aristotle's purpose to show us *how* he arrives at the correct definition of "animal," nevertheless the definition that he proposes is correct, because it correctly orders all the essential predicates of "animal." These essential predicates that define the animal are the sensory, digestive, and locomotive systems (3). Aubonnet remarks, in a note to his translation, that this list is the most complete that one finds in Aristotle.[10] I must take a further step and assert that the definition is exhaustive. Self-nourishment is, in fact, the minimum criterion for life, as is stated in several places, including this passage from the *De Anima:*

> Among natural things, some possess life, and others do not; I call life the activity of nourishing oneself, growth, and decline.

> *(2.1.412a13)*

This is just one criterion, since growth and decline are only the most obvious corollary of nutrition; furthermore, this is confirmed by the Aristotelian teaching concerning the growth of living things. On the other hand, animal life is characterized by sensation and motility, as Aristotle indicates in many passages where he opposes animals to plants.[11] In my opinion, moreover, the first part of phrase 4 should *not* be taken as a contrary-to-fact conditional, as both the French translators cited above have done — in any case, that would require an ἄν in the following clause — but means: "and if these parts are indeed the only ones" (i.e., are necessary for defining the animal), with "as is indeed the case" understood. This passage is not primarily biological, but uses zoology only as the basis of a comparison; it presents the results of Peripatetic biological research without having to justify those results. Similarly, Aristotle had already alluded to the results of logical research, the diairetic construction of definition, without directly employing *diairesis.* However, we must recognize that the definition of "animal" given here is defective on one point, for it does not give its constitutive predicates in the order recommended as adequate in the *Analytics:* self-nourishment, the most general predicate, should have come first, then sensation, common to all living animals, then motility, which denotes a superior degree of organization and which thus tends to disappear in zoophytes.

It is only then, after phrase 4, that the process of division of the *genos* by specific difference begins. However, this process is not applied to the *genos* "animal," but to *gene* of essential organs (5), which have *diaforai.* Here again, we find the same method that we noticed when we discussed the beginning of the *History of Animals* (1.1.486a14). In fact, that method is fundamental for all Aristotelian biology: the biologist should explicate the combination of *parts,* since it is at this level that the world of living things is revealed to be truly intelligible. For as we have seen and will see again, the diversity of animals is only a consequence of the diversity of the parts. It is incontestable that in the biological works the study of digestive, sensory, and locomotive organs is indeed carried on as the study of *eide* within *gene,* the identity of function thus delimiting the generic unity. And we do indeed find, within each *genos,* the play of contrary *eide;* this is so in the case of the belly,[12] concerning which Aristotle writes in the *Parts of Animals:*

> We must now examine the specific difference *(diafora)* of the belly and adjoining parts; for not all the various animals have bellies that are similar in either size or "form" *(eidos).*   *(3.14.674a21)*

Here the topic is clearly the operation of specific difference *(diafora* is used in the singular) within the *genos* "belly." The difference in respect of *eide,* mentioned at the end of the passage, might be understood as a difference of configuration. That is how the translators I have consulted render it, except perhaps the ambiguous Didot Latin, "neque specie similes." It could also be understood as an indication that the *genos* "belly" has several *eide.* I am inclined to accept this interpretation, understanding that Aristotle here puts side by side, the one reinforcing the other, quantitative and specific differences.

The *History of Animals* is more precise:

> All animals have common parts by which they absorb and within which they keep their food. These parts are similar and different in the ways we have specified: in respect to *eidos,* excess, analogy, position.   *(1.1.488b29)*

This passage shows more exactly that there are several *gene* of organs within one single function: thus, for the nutritive function, one may distinguish mouth and belly; and each of these organs is divided into *eide.* The fact that certain organs belong to the same *genos* is not always immediately obvious. In the ascidians *(tethya),* despite their "most extraordinary nature" *(HA* 4.6.531a8), thanks to scientific investigation (dissection, 531a16) we find an organ that must be accepted as belonging to the *genos* "mouth": the ascidians have, in fact, a passage "which is such that one may take it as if it were a mouth" (531a23).[13] It would be easy to show the same thing in the case of other fundamental functions.[14]

At this most general level, in fact, the *gene* of digestive, sensitive, and locomotive organs no longer have between them an analogical relationship, for in what sense could one say that the mouth is the analogue of the foot? The only possible relation between them is their "conjunction" (phrase 6) or "combination" (phrase 8) in "ultimate species." And, in any case, the doctrine of organic analogy between animals of different *gene* rests on the generic distinction of functions: it is by presupposing a *genos* of "locomotive organs" that one may say that "wings are to birds as fins are to aquatic animals." In fact, analogy directly implies a combination of the same and other: in the case of wing/fin, for example, the otherness is immediately perceptible, for wings are not fins; their similarity flows from their both belonging to the *genos* "locomotive organ." Does this mean that if the fundamental functions were so distant from each other that there would no longer be any analogy between them, they would then no longer have any relation at all? There does, in fact, exist a "bracket" that unites these three *gene* of basic functional parts: that is life.[15] As I have remarked above, we may therefore range these three functions in order of increasing determination. We thus get this table:

$$
\text{life} \begin{cases} \text{self-nourishment: plants}^{16} \\ \left. \begin{array}{l} \text{sensitivity} \\ \text{motility} \end{array} \right\} \text{animals} \begin{cases} \text{sensitivity only: inferior animals} \\ \text{sensitivity + motility: perfect} \\ \qquad\qquad\qquad\qquad\quad \text{animals}^{17} \end{cases} \end{cases}
$$

This order is necessary: each faculty other than the first is present potentially in the one that precedes it:

> Both in figures and in things that possess soul, the earlier always exists potentially in what follows; for example, the triangle is included in the quadrilateral, and the nutritive faculty in the sensitive. Consequently, for each being we must see which soul belongs to it: for example, what is the soul of the plant, the man, the animal.
>
> (De An. *2.3.414b29*)

That self-nourishment is the widest predicate is abundantly proven by the references given above. That sensitivity is more "primitive" and general than motility is shown by the rising ladder of perfection of animals developed by Aristotle, but one may also find its basis in his general theory of movement. For Aristotle, locomotion is prior except in relation to beings subject to generation:

> Secondly, locomotion must be primary in time: for this is the only motion possible for eternal things. It is true indeed that, in the case of any individual thing that has a becoming, locomotion must be the last of its motions: for after its becoming it first experiences alteration and increase, and locomotion is a motion that belongs to such things only when they are perfected.
>
> (Physics *8.7.260b29, trans. Hardie and Gaye*)

If we remember that for Aristotle, on the one hand, sensation is an alteration (e.g., *De An.* 2.4.415b24) and that, on the other, growth is an effect of nutrition, we find in this passage of the *Physics* the three fundamental vital functions.

There is here, if I may say so, a "relation to one unique term" *(pros hen)*:[18] all these functions can be related to one unique term that includes them all, and that is life. In this sense, for the person who considers the world of living things, life is rather comparable to being for the person who occupies himself with the science of being *qua* being: life, the one term in relation to which all vital phenomena are determined, is itself determined in several ways, and the fundamental functions that we have enumerated would be, as it were, "categories of life," just as the various ways of determining being are the categories of being. Perhaps Hegel had a notion of this sort of interpretation when he wrote: "Aristotle's conception of nature is, however, nobler than that of today, for with him the principal point is the determination of end as the inward determinateness of natural things. Thus he comprehended nature as life." [19] Nevertheless, we must avoid seeing in this relationship between biology and ontology more than a comparison; it is especially certain that there is no analogy—in the strictly Aristotelian sense of the word—linking the relations of substance, the fundamental category of being, to the other categories, and the relations of self-nourishment, the fundamental function of life, to the other vital functions.

Nevertheless, one may suppose that this Aristotelian "vitalism" puts still more distance between Aristotle and the taxonomic point of view. In fact, Michel Foucault (1966, 173) was right when he said about the great period of taxonomy: "Natural history, in the Classical period, cannot be established as biology. Up to the end of the eighteenth century, in fact, life does not exist: only living beings. . . . [A]nd if it is possible to speak of life, it is only as of one character — in the taxonomic sense of that word — in the universal distribution of beings." Aristotle, in contrast, is, in the etymological sense of the word, a *biologist:* life is not a predicate,[20] but living things manifest life, each one in its own place. It is thus that, beneath their phenomenal diversity, they are profoundly the same, according to the formula already cited from the *History of Animals* (1.1.486b23). The Aristotelian doctrine of the structural and functional unity of living things translates their deeply rooted identity.

To return to the passage from the *Politics,* when Aristotle finally asserts that there are as many animal *eide* as there are possible combinations of the various fundamental organs (8), we may see in retrospect how distant his procedure was from that of the taxonomists. This Aristotelian method, which permits the discovery of reality by the *a priori* construction of the possible, cannot be that of a systematician: one cannot find even a distant relationship between it and the botanical method of Linnaeus, for example, described in summary above. If we really must find its renaissance, we would have to look for it in Mendeleev,[21] or before him, in a sense (as we shall see), in Cuvier.[22]

I must make one last remark about this passage in regard to the word "necessarily" in phrase 6. It is used there in a logical — in fact, quasi-mathematical — sense, since if we have a given number of parts, and their possible conjunctions, we necessarily (mathematically) derive a number of possible kinds of animal.

Aristotle is able to assert, on the ground of the principle that phrase 7 poses, that there are *necessarily* as many species of animals as there are combinations of organs: the same animal cannot have several organs of the same kind. But that rests, in fact, upon a teleological principle, since it is ultimately derived from that other principle, so many times asserted, that nature does nothing in vain and always brings about the best.[23]

Even in this passage, then, although it has the reputation of being "deductive" and "classificatory"—which is essentially the position of G. E. R. Lloyd in the article examined above in Chapter 1—Aristotle's biology remains *radically* teleological. That is why, if we want to appreciate exactly the status and function of the animal classifications that Aristotle constructs, we must put them into the framework of the four causes, dominated by the final cause, proper to the Aristotelian philosophy of nature.

Having definitively established that there is in Aristotle no shadow of a taxonomic project, if we wish to investigate the functions of Aristotle's classifications in zoology, we must redefine, at least partially, the status and methods of Aristotelian zoology. The examination of that status will lead us to try to grasp the logical structure, in the various treatises, of Aristotle's biological research. Then, and only then, will the place and functions of his animal classifications be clear.

Let us remember, in the first place, that for Aristotle zoology is a part of "physics." Animals are, in fact, natural entities. Notice that they are always, or almost always, the first examples given of "natural" beings—for example, at the very beginning of *Physics* 2, in which Aristotle explains the idea of *physis:*

> Of things that exist, some exist by nature, some from other causes. "By nature" the animals and their parts exist, and the plants and the simple bodies (earth, fire, air, water)—for we say that these and the like exist "by nature."   *(2.1.192b8, trans. Hardie and Gaye)*

In this passage, Aristotle seems to join philosophy and common sense, since these opinions are shared by all or most people. Within the biological treatises, too, biology is said to be a part of "physics." Thus, at the very beginning of the methodological *Parts of Animals* 1, zoological research is included in the study *(historia)* of nature (1.639a12). Consequently, zoology, as a part of "physics," is a *theoretical* science, according to the traditional Aristotelian classification of the sciences.[24] Thus, I must agree with the interpretation that Ingemar Düring (who followed Michael of Ephesus, in contrast to most modern commentators) provided of a passage of this first book of the *Parts of Animals,* translated thus by Peck:

> Howbeit, the method of reasoning in Natural Science and also the mode of Necessity itself is not the same as in the Theoretical Sciences.[25]

Several editors have noticed the difficulty of this passage, for it seems to contradict Aristotle's usual classification of the sciences. Thus, Peck suggests that we introduce this distinction: "Our study of Nature's science may be a 'theoretical' science, but Nature's science itself is 'productive'." [26] In his edition of Book 1 of the *Parts of Animals,* Jean-Marie LeBlond speaks of *"flottements"* of Aristotle's terminology, and, explicitly rejecting Michael's interpretation, cites two other passages in which he thinks he finds a doctrine that conforms to his translation.[27] Düring's reading of this passage (1943, 216) allows us to avoid these terminological vacillations:

> As it stands, the text means: "Yet in physical science, as well as in the theoretical sciences, the method of demonstration and the mode of necessity are different." Michael interpreted *héteros* as "different from the one we have just discussed," i.e., "different from the method of demonstration (reasoning) and the mode of necessity in constructive science." [28]

In fact, in the sentence that precedes our passage, Aristotle has just given the example of the necessity that regulates the construction of a house in order to illustrate the difference between simple and conditional necessity:

> One must have at hand such and such material in order to actualize a house or some other end: and a determinate thing must be produced and put into movement first, then some other specific thing, and so on until the end, i.e., that for which each thing is produced and is. It is the same in the case of things that come to be naturally.
>
> *(639b26)*

Then comes our passage, which opposes to this necessity of fabricated entities the necessity that occurs in natural and theoretical sciences. We must take the conjunction to mean: in the natural sciences, and more generally in the theoretical sciences. Düring goes on (1943, 216):

> The theoretical sciences are geometry and other branches of mathematics; physical science, is, on the one hand, that kind of speculative physics that we know from the eight books of the *Physics;* on the other hand, and especially so in the present discussion, the science that we call biology. In all the fields of science, Aristotle says, we begin our investigation and reasoning with what exists, τὸ ὄν. In the constructive arts, in building a house or making a statue, we begin with an idea of the result to be achieved, τὸ ἐσόμενον. As a whole, this sentence is an answer to the question raised at 639b7: "Should the φυσικός follow the same method as the mathematicians?" And Aristotle's answer is "Yes."

If zoology is indeed part of "physics," and is thus theoretical, it searches for the causes and is concerned with what is and not with what ought to be. Thus, Peck translates the sentence that immediately follows our passage:

> (I have spoken of this matter in another treatise.) They differ in the following way. In the Theoretical sciences, we begin with what already *is;* but in the Natural sciences with what *is going to be.*
>
> *(640a2)*

Thus, according to Peck (and Louis, and Balme), the "speculative sciences" are concerned with "what is," and the "natural sciences" are concerned with "what is going to be." This reading seems at first glance to be self-evidently true, since animals are indeed subject to generation, that is, to becoming. Furthermore, other passages (*Posterior Analytics* 2.19.100a9; *Nicomachean Ethics* 6.4.1140a10) tell us that "being" (τὸ ὄν) is to *genesis* what *episteme* is to *techne*. Nevertheless, Düring's interpretation is correct; the examples that we have noted are sufficient to show that Aristotle here means by "what ought to be" an attribute of "things that can be other than they are": in the passage from the *Parts of Animals,* he speaks of a house; in the two other passages, of productions of art. This *genesis* of objects of art cannot be the same as the generation of animals: when we recognize the importance to Aristotle of the opposition between manufactured objects and natural beings, we see that he takes them as two different domains. Objects of art can be other than they are because they depend on the vagaries of human intention and capacity. Aristotle insists that the necessity found in the development of every animal is *internal*. From the point of view of necessity, animals are closer to the everlasting beings than to artificial objects, since they are beings that cannot be other than they are. That is shown by what we have already seen concerning the *genos* as the place in which the *genesis* of living things occurs; by attributing to them a kind of generic everlastingness and necessity, *genesis* gives them the status of objects of science. Aristotle several times repeats that the cycle of generations is an imitation of cosmic everlastingness (e.g., *De Anima* 2.4.415a26). In his attack on Empedocles' notion that the curve of the spine occurred subsequently to the generation of the animal, Aristotle says that this curve exists potentially in the seed of the animal (*PA* 1.1.640a19). Clearly, the living being is not constructed like a house.

It is very important for our present study that zoology is indeed both scientific and theoretical for Aristotle, and yet we shall see that the arrangement of living things into families could not be a scientific (epistemic) procedure. Gilles-Gaston Granger (1976, 14ff.) characterizes science, in the Aristotelian sense of *episteme,* by three axes that he calls psychological, phenomenological, and logical. He then approaches science from several sides and defines it as: *doxic* in that, in contrast to imagination, *episteme* affirms and denies; *discursive* in that, contrary to opinion, it claims to base its assertions on a demonstrative discourse; *predicative* in that, contrary to perception and all forms of immediate cognition, it combines or separates concepts; *symbolic* in that, in contrast to perception, it abstracts its objects from their immediate existence. Furthermore, it concerns itself with necessary objects; and, as the *Nicomachean Ethics*

(6.3.1139b25) says, "every *episteme* is communicable by teaching." Granger summarizes (p. 24): "Science is thus ultimately characterized by the capacity of being transmitted, by the necessity and eternity of its object, reflected in the necessary chain of propositions that describe it, and by the presentation of the causes."

But the grouping of animals into families is neither apodictic nor etiological: it is a recognition and an ordering. We must locate an operation of this kind on the psychological axis, the one of the three that interests Granger least. On the psychological axis, "the process of thought is taken as a natural event, having a genesis, causes, effects" (p. 12). If one takes as a conceptual frame the continuist model[29] of ascent toward scientific knowledge, presented to us at the beginning of the *Metaphysics,* zoological classifications belong to experience *(empeiria),* or, at most, to art *(techne):* as a first sorting of the immediate given, those classifications resemble art in that they take hold of a universal, but they resemble experience in that they cannot achieve knowledge of the cause. Or, to use another Aristotelian distinction, these classifications are on the level of the fact *(hoti)* and not of the why *(dioti).*[30] In contrast, there is a biological science: apodictic and etiological research carried on according to the four causes. I shall try to show that Aristotle's animal classifications are a prologue to his scientific research in zoology by partially summarizing his actual procedure in that research. Although the classifications cannot claim scientific rigor, they help to prevent scientific research from getting lost in the pure multiplicity of the immediate.

If Aristotle's zoological procedure has been almost completely neglected by previous commentators, it is perhaps because, among other things, they have been fascinated with chronological problems. I must say a few words about this before getting started. Discussions of the chronology of the biological corpus have two aspects: questions concerning the external chronology ask about the place of this corpus in the unfolding of Aristotle's intellectual production, whereas questions concerning the internal chronology aim at settling the order of composition of the various biological treatises. Regarding the first point, in Chapter 1 (note 26), I quickly summarized the opposed positions of Werner Jaeger and D'Arcy Thompson. On the second point, which interests me more here, there has been an almost universal agreement that the chronological order is: *History of Animals, Parts of Animals, Generation of Animals.*[31] David Balme has carried out detailed studies on the biological corpus, especially the *History of Animals,* during the past several years, and has become convinced, on the contrary, that this work was the last composed by Aristotle.[32] These chronological studies, though certainly interesting, suffer from a fundamental defect: unable to rely upon objective criteria, that is, stylometric criteria, they rely ultimately on doctrinal considerations. All interpretive orders are thus possible, and all are ultimately determined by the ideological positions of the interpreters. That is especially obvious in the two examples cited of external chronology: what ultimately determines the disagreement between Jaeger and Thompson is that they do not have the same concept of the relationships between philosophical speculation and empirical research.

As for internal chronology, I shall limit myself to the examination of the thesis that places the *History of Animals* prior to the other treatises. My critiques of that position are not of the same nature as Balme's. Balme's two principal arguments in favor of the posteriority of the *History of Animals* are: (1) that when this treatise reports the same facts as the others, it is generally briefer than they are, and thus the *History of Animals* summarizes results already presented elsewhere; and (2) that among the facts that only the *History of Animals* reports, many contradict assertions made in the other treatises, which would allow one to suppose that these facts have been acquired after the writing of the other treatises. If I am satisfied with only a rather rough summary of Balme's theses, it is because my critique starts from a quite different standpoint. I would like, in the first place, to ask about the ground of the thesis, generally accepted before Balme, that the *History of Animals* is the first treatise in the biological corpus. Partisans of that thesis generally rely on Aristotle's own assertions and on internal references in the biological writings. I shall speak about those in a moment. But I believe that the thesis really rests on anachronistic presuppositions, of which two appear to be primary.

In the first place, as the *History of Animals* presents "more" facts and "less" explanation, the modern reader, formed, whether he likes it or not, in the conceptual framework of experimental science, will be almost irresistibly led to think that the collection of facts *ought* to precede the explanation of those facts. I shall show in a moment, in relation to one precise example, that of the heart, that for Aristotle it is not at all necessary that observation precede the explanation of phenomena. The second presupposition is related to the very form of the biological treatises. Interpreters cannot help thinking of the biological treatises as "books" in the modern sense of the word, even though they are quite aware of the doubtless collective character of biological research in the Lyceum and of the fact that the Aristotelian corpus has come down to us in the form of unedited notes. At most, interpreters concede that the *History of Animals,* for example, may be considered as a group of several books. I hope to show that if one takes seriously the evidence we have concerning the way in which the *History of Animals* was composed, the idea of placing this treatise either before or after the *Parts of Animals* is just simply senseless.

However, Aristotle does explicitly clarify the interrelationships between the three large treatises of the biological corpus in several places. These passages are, for the most part, transitional or introductory; Aristotle notes in them how far his research has gone and where it will go, and it is these passages that have been used by interpreters to support the thesis that the chronological order is *History of Animals, Parts of Animals, Generation of Animals.* We must look at several of these passages, beginning with one that seems, at first sight, to contradict the others.

In the first book of the *Parts of Animals,* a book thought to be "methodological," Aristotle posits a basic question:

Is it the case that, like the mathematicians in their astronomical research, the natural scientist ought in the first place to consider the facts *(ta fainomena)* relative to animals, and the parts of each of them, and then give the why and the causes, or ought he to proceed otherwise? *(1.639b6)*

As Düring has shown, Aristotle thinks that the naturalist should proceed as the mathematician does.[33] When he takes up this question again a little later, Aristotle enriches his research project with a supplementary stage:

We must first grasp the facts about each kind of animal, then present the causes, and finally the generation. *(640a14)*

According to this passage, there would be three stages in biological research: phenomena, causes, generation. The partisans of the traditional order see in this an indication of the order of the treatises: "that is the order that the corpus as a whole follows," writes LeBlond (1945, 140, n. 29), commenting on this passage.

But most of the transitional passages in Aristotle's biological work organize the project differently, according to the theory of the four causes. Let me quote two such passages, both pivotal, for one opens Book 2 of the *Parts of Animals,* that is, the biological treatise itself, after the "discourse on method" in Book 1, and the other is found at the very beginning of the *Generation of Animals:*

The *History of Animals* has shown clearly enough from which and from how many parts each of the animals is composed; now we must examine the causes of each animal having such a characteristic.

(PA 2.1.646a8)

Thus, we have spoken about the other parts of animals in general and in particular, including their special attributes, and about the way in which each exists according to a determinate cause that I call final cause. In fact, there are four causes: that for the sake of which, which is the end, and the definition of the entity. These two causes can be considered as forming effectively only one. The third and fourth are the matter and the source of movement. Since we have already spoken about the others (since the definition and the end for the sake of which are the same thing, and the matter for animals is their parts, for the animal as a whole the anhomoiomerous parts, for

the anhomoiomerous parts the homoiomerous parts, and for these
what we call the elements of bodies), it thus remains to treat of those
parts that contribute to the generation of animals, to which we have
not yet devoted a special study, and it remains to say what is the
motive cause. The study of this cause and that of the generation of
each animal is in a way the same thing.   *(GA 1.1.715a1)*

I have two points to make about these three passages. In the
first place, they are not mutually contradictory; they propose
plans for the study of living things that are different but can be
superimposed. The first passage is in fact *methodological,* in that it
guides the searcher struggling with the biological given, while
the other two passages are *systematic,* in that they indicate what is
necessary and sufficient for the Aristotelian scientific "contract"
to be fulfilled. An object is not adequately known except when it
has been studied from the point of view of all four causes, even if
in the case of animals Aristotle says that the formal and final
causes are more or less identical. Perhaps it is utterly impossible
to know whether these programmatic passages were or were not
added after the composition of the treatises they introduce; but
either way, the Aristotelian biological corpus is still, so to speak,
"framed" by an explicit and systematic research project based on
the doctrine of the four causes.

Nevertheless, and this is my second point, this systematic
order does not necessarily reflect a chronological order. By giving each of these treatises a specific etiological goal, Aristotle
establishes a theoretical division of labor between them and
simultaneously indicates a systematic order of reading — or of
presenting — the biological corpus. Nothing authorizes us,
however, to draw chronological conclusions. Balme was quite
right when he said:

It has usually been taken for granted that the *HA* came earlier,
because in it Aristotle says that it is to contain the investigation of
facts which must precede the investigation of causes, and this procedure agrees with what he says in the *An. Post* and *PA* 1. But this
statement comes in the introduction to *HA* and clearly refers to the
order of lecturing.[34]

I shall try to clarify this systematic division of labor in etiological research by commenting on each of the large treatises of the biological corpus.

The *History of Animals* is immediately opposed to the other treatises, *Parts of Animals, Generation of Animals, Movement of Animals,* and *Progression of Animals,* by special characteristics that cannot be denied. But let us try to avoid being fooled by that difference. Much longer than the other books, the *History of Animals* offers a great deal more information than all the other treatises put together. Many animal characteristics are noted there but ignored by the other treatises when they touch on the same topics. Many kinds of animals appear only in the *History.* The viewpoints for considering those animals are most varied: parts and habits of animals, generation and nutrition, husbandry and veterinary medicine, and so on. However, the *History of Animals* is not at all a gathering of disparate observations, as would be many of the zoological compilations after Aelian until the sixteenth century. Today, no specialist would support the notions of Georges Pouchet, who said (1884, 362): "It is really in vain that some have tried to discover a plan in this incoherent mass of most diverse topics, in this book that seems made of the debris of an entire library from which one has saved several volumes from random shelves and others from forgotten corners."

This very abundance of data has been one of the major causes for leading interpreters to think of the *History of Animals* as a storehouse of observations available for later scientific research. That is the position of, among others, LeBlond.[35] Other facts seem to tend in the same direction. Thus, Aristotle himself several times alludes to anatomical drawings or descriptions that must have accompanied the treatise.[36] This character of the *History of Animals* as a "reference" work is incontestable; Franz Dirlmeier (1962) has demonstrated this conclusively in a remarkable study. He begins with a general examination of the status of writing and its relationships with oral presentation in

Aristotle, and finds that only the *History of Animals* is cited, in the other works, with the verb "to write" *(grafein)*. For example, in the treatise *On Respiration* (477a5): "This topic *has been written* about more precisely in the *History of Animals.*" Dirlmeier concludes from this that the *History of Animals* was a written reference work *(Nachschlagewerk)* that Aristotle's hearers could consult.[37]

On the basis of all these arguments, we can certainly accept the idea that the *History of Animals* was indeed a gathering of observations, meant to serve as a reference. But I believe that a further step, this one unacceptable, is taken by the commentators when they conclude that the *History of Animals* was *only* a reference work. This treatise would then have the same pre-scientific status as the various classifications of animals — that of a study preliminary to properly scientific speculation, which is the search for causes. Auguste Mansion is thus in error when he writes (1945, 24): "Here we distinguish the ὑπομνήματα from the didactic treatises. The first are collections of facts, compiled largely by Aristotle himself, but parts could have been entrusted to students. . . . We see immediately that we must place the well-organized gathering or collection of facts which is the *History of Animals* into the class of ὑπομνήματα." As we have seen, this was also Jaeger's position. And some of Aristotle's passages themselves seem to tend in this direction. Thus, at the beginning of the *Progression of Animals,* he lists the problems to be examined: the number of points at which animals move, why blooded animals have just four, why some animals are bipeds and others quadrupeds, and still others many-footed, and so on. The research program thus proposed is clearly etiological. Then Aristotle writes:

> We should try to discover the explanations of all these facts, and of related matters. That this is what actually occurs is made clear in the *Natural History;* now we shall find out why.   *(1.704b8)*

I would, however, like to show that this idea of the *History of Animals* as a reference work is not only insufficient, but it obscures what I have called the internal structure of Aristotelian zoology. To this end, I shall analyze the famous passage following that concerning the "very large *gene*," which we discussed in Chapter 2:

> That is why we must study [the various kinds of animals] by examining the nature of each one separately. What has preceded is now considered only a sketch, a foretaste of the questions and objects to be studied. In what follows, we shall deal with them with precision, in order to grasp, in the first place, the differences and the common attributes of all the animals; after that, we should try to discover their causes. . . . [By this method] will appear more clearly the object and the premises *(ex hon)* from which there should be demonstration *(apodeixis)*. In the first place, we should consider the parts of which animals are composed. For it is about them that the principal and first differences of the animals considered as wholes show themselves. (HA *1.6.491a4 et seq.*)

I shall not discuss the question of the range of animals referred to by the first sentence — all animals or only the viviparous quadrupeds that he has just been talking about — for that is unimportant to us here. In this passage, Aristotle distinguishes two stages in his research. First, there is "a sketch, a foretaste"; then there is a "precise" study, which is itself divided into two stages: first, of "differences and common attributes"; then, of "causes." Nowadays, interpreters readily agree about two points: first, that the "sketch" of which Aristotle speaks is a preliminary study and is not part of the scientific research itself; thus, it may be called "prescientific"; and second, that the "sketch" is not the study undertaken in the *History of Animals,* for no one could argue that this book is a collection of studies of individual kinds of animals.[38] As for the division of the "precise" research — on the one hand, "differences and common attributes," and on the other, "causes"— can we suppose that it reflects the distinction between the *History of Animals* and the other biological treatises? That is the position of, among others, Pierre Louis and David Balme. Louis writes: "The first point will be the object of the present treatise [i.e., the *History of Animals*], the second that of the special treatises."[39] As for Balme, he comments on this passage (1975, 192):

> Yet Aristotle does state his purpose: "first, to grasp the differentiae and attributes that belong to all animals; then to discover their causes" (*HA* 1.491a9). The *HA* is a collection and preliminary analysis of the *differences* between animals. The animals are called in as witnesses to differentiae, not in order to be described as animals.

I believe that Balme is right: the principal objective of the *History of Animals* is the study of the differences that exist between the parts of animals, even if the *History of Animals* is not only concerned with that, and indeed, according to a passage already cited, treats the "differences between animals according to their mode of life, their actions, their character, their parts" (1.1.487a10). Let me only remind the reader of the problem of the extent of the original treatise, the "primitive" *History of Animals*. Must we think, as Düring does, that there was a first edition of the treatise that was limited to the first six books of the *History* as we know it?[40] That would tend to support the thesis that the *History of Animals* has an essentially "moriological" (part-studying) character, but it seems to me too uncertain to use this as a true proof.[41]

However, this passage does not at all support the interpretation according to which the *History of Animals* would only present the data that the other treatises would explain with the help of the theory of the four causes, for under that interpretation one of the most important phrases in the passage would be unintelligible: "[By this method] will appear more clearly the object and the premises from which there should be demonstration" (491a12 – 14). I believe that here we must understand the word *apodeixis* in its technical sense of "demonstration," and not in the vague sense of "exposition," as Peck, for example, takes it. (Thompson's "argument" is better.) Aristotelian zoology is both demonstrative *(apodictic)* and teleological; we shall return to this point in our study of the other treatises of the biological corpus. Thus, when we have established that bile — or at least a part of the bile — comes from the liver and that it is a residue, we understand why bile not only has no purpose of its own, but is even something pathological resulting from a bad functioning of the liver. Its absence would then be evidence for the good functioning of the liver, and therefore of long life. Thus the proposition that "animals with no bile live long," which in its turn serves as a premise for several syllogisms.[42] But only this systematic review of the *differences* concerning bile — bile coming from the liver or the intestines, a difference of secretion from one species to another, and, more difficult to observe, from one individual to another within the same species — will permit the establishment of universal propositions.[43]

It seems to me that by assigning to the *History of Animals* the function of establishing the premises for demonstrations, Aristotle makes this treatise an integral part of his etiological research program. More precisely, the *History of Animals* is characterized by its own etiological goal: it is charged with studying living things from the point of view of "material causality." Aristotle's material causality is much wider than the picture that commentators generally give it, for they have, more or less consciously, been influenced by the modern sense of the word "matter." Let us consider two passages. In the first, taken from the *Posterior Analytics,* Aristotle, enumerating the four causes, says of the "material cause": "For some things to be, something else must be." [44] As we know, that defines the syllogistic relationship. The other passage is found both in the *Physics* (2.3.195a16) and in the *Metaphysics* (Delta 2.1013b17):

> Letters for syllables, matter *(hyle)* for artificial objects, fire and the other elements of that kind for bodies, parts for the whole, premises *(hypotheseis)* for the conclusion, are causes as that of which the things are made.

Granger writes (1976, 117): "It would seem, then, that the causality expressed by the syllogistic relationship is identified with 'material' causation." Thus, it seems that the relationship that we think of as the most "formal" (in the modern logical sense), that of premises to conclusion, is for Aristotle a material relationship.

The passage from the *History of Animals* analyzed above (1.6.491a4 et seq.) and the opening section of the *Generation of Animals,* already quoted, converge toward the same doctrine. When the latter passage says that "we have spoken about the other causes, . . . since matter for the animals is their parts" (715a9), it indeed seems understood that this study was carried out in the *History of Animals,* especially if one compares this passage with that at the beginning of Book 2 of the *Parts of Animals,* which I quoted at the same time. The *Generation of Animals,* the last major treatise of the Aristotelian biological corpus — if not chronologically, then logically — thus opens with a methodological passage locating each treatise within a general theoretical project in which it participates in respect to its own etiological goal. Whether this methodological passage was or was not added after the fact does not matter much to us, since we are not here concerned with a chronological issue. In this division of labor, the *History of Animals* seems to be the *first* but not the *preliminary* study, and the nuance is important because it means that the *History of Animals* is a "scientific" treatise in its own right, characterized by its own etiological goal.[45]

In summary, I believe that the interpreters who claim that the *History of Animals* is a collection of data and observations are not mistaken. An additional piece of evidence, besides what we have already adduced, would be that either Aristotle or his later editors found it convenient to include many "raw" observations in the *History of Animals,* observations without any explanation, particularly about animal ethology. But this thesis is one-sided and inadequate; we must reject the excessive theoretical devaluation of the *History of Animals* characteristic of most interpreters. Aristotle wanted (either from the start or when he later resystematized, it doesn't matter which) to include the *History of Animals* as a part of his scientific, or etiological, project. The theoretical heart of this treatise is to be a study of the parts of animals via their differences. According to ancient writers, particularly Athenaeus, we should therefore read this treatise as dealing with the parts of animals. It is quite remarkable, from our point of view, that the *History of Animals* thus "rehabilitated" gives no space to any explicitly formulated taxonomic project. This absence already indicates something that a study of the *Parts of Animals* will establish even more firmly: namely, that for Aristotle the various orderings of animals are not part of his *scientific* research.

The *Parts of Animals* studies what Aristotle calls, in the passages cited and in many others, "the causes," that is, the formal and final cause, the causes *par excellence* in the biological domain. The first thing that strikes anyone who compares this treatise with the *History of Animals* is that it introduces very little new content: only description of the parts and habits of animals. But these books differ in perspective: whereas the *History of Animals* observes and describes, the *Parts of Animals* explains. Thus, the *History of Animals* notes (or claims to note, for it is a surprising observational error on Aristotle's part) that the human male has more sutures in the skull than the human female (1.7.491b2); and the *Parts of Animals* explains this supposed fact by the larger volume of the masculine brain (2.7.653a37). In my opinion, Louis Bourgey misunderstood the real difference between the two treatises, which is one of viewpoint, and so he writes concerning the heart (1955, 86, n. 6):

> A first description of the cavities of the heart is given in the *Research on Animals* 1.17.496a4–27; this description corresponds very precisely to that in the *Parts,* but the final cause that justifies the separation of the three cavities is not at all indicated, a manifest proof that this cause was found after the fact.

Between the two treatises there is a division of labor unrecognized by Bourgey: the description is the study of the material cause. Even if we have a tendency to think of the description as earlier, we cannot be sure: the two etiological goals, material and final, can very well be contemporary, and nothing prevents us from thinking that Aristotle's procedure may have been more nearly that of looking for observations to support an idea that was first developed *a priori*. It is possible to argue that Aristotle most probably did work this way: consider his observational error in "finding" only three cavities in the heart. It is possible that the observation came *after* the teleological reasoning and that it was hypothesized and, as it were, obscured by the metaphysical necessity of having a single center in a higher animal.

> In the large animals, the heart has three cavities, in the smaller ones, two only; and in no species has it less than one. . . . Now, there are two chief blood vessels, the so-called Great Blood Vessel and the Aorta; each of these is the source of other blood vessels; and the two differ from each other . . . ; hence, it is better for them to have separate sources. This result can be obtained by having two separate supplies of blood, and thus we find two receptacles wherever this is possible, as in the larger animals, which of course have large hearts. But it is better still to have three cavities, and then there is an odd one in the middle that can be a common source *(arche)* for the other two.   *(PA 3.4.666b21)*

In this case, the teleological idea that nature always accomplishes the best possible has contributed to preventing Aristotle from observing correctly. There is more than one example of this kind in his biological work.

The *Parts of Animals* thus constitutes the central treatise of Aristotle's zoology, and its preeminence flows from the preeminence of the final cause. To see the place and functions that are assigned there to animal classifications, we must first clarify the purpose of this treatise.

No one can doubt that the work that we know under the title *Parts of Animals* is composed of two parts that differ greatly from each other: one part is Book 1, a methodological exposition; the other is the remaining three books, to which alone the title really applies. This division surely does not justify the removal of Book 1 from the "biological writings"; Balme was wrong to do that in his study (1962a) of the terms *genos* and *eidos* in Aristotle's biology. Nevertheless, I shall speak here of Books 2, 3, and 4 as having a unity of means and purpose. Their objective is to examine the "manner of being" *(tropos,* 2.1.646a10) of each of the parts spoken about in the *History of Animals* — that is, the point is to demonstrate that what we would call the structural

properties of these parts — from the configuration of such and
such an organ to the consistency of such and such a humor — are
determined by the functions of each of these parts, and from
among all the possibilities nature chooses always or most often
the best. Besides this teleological principle, this demonstration
rests on three others of the same kind: nature does nothing in
vain, it does not give on one side without taking something
away from another, and every living thing is adapted to the
environment in which nature has decided to put it.

Thus, Aristotle tries to arrive at universal correlations, which
may be taken as true universal premises. For example: "All
living things that have a lung also have a neck" (4.10.686a3). It
would seem that the establishment of such correlations, to the
degree that they are really universal, requires the identification
and study of various animal kinds. We have seen that this preliminary work was thought necessary in the plan presented in
the *History of Animals.* That poses for Aristotelian research the
problem of the distribution of animals into distinct groups.
Thus, we must look more closely at an important question: did
Aristotle, with the help of his school, attempt to list all the
species available at the time? If he had that project, one might
imagine that he was trying to construct animal classes *inductively.* Furthermore, one might imagine that since he thought
this construction prescientific (in that it was not etiological), he
could not include it in the scientific treatises, and that we have
for this reason lost all trace of his inductive constructions. Indeed, one may get the impression from the *Parts of Animals* that
Aristotle was using a classification of animals that he took to be
already established, a classification resembling that which most
commentators have taken to be *the* Aristotelian classification of
animals. Among animals with blood, he distinguishes the viviparous quadrupeds (among which he sometimes includes man),
the oviparous quadrupeds (sometimes including serpents), birds,
and fish; among the bloodless animals, he distinguishes molluscs, testaceans, crustaceans, and insects. One might naturally
suppose that Aristotle would have taken every precaution in

assembling the most exhaustive possible census of species, lest an unexpected exception ruin the universality of a correlation and thus its scientific value.

But despite Aristotle's encyclopedism, we find no such attempt at an exhaustive census either among the works that have come down to us or even in the ancient lists of Aristotle's works. One may well imagine that such a gigantic enterprise would have left at least some trace in the memory of the doxographers; and the prescientific character, in Aristotle's eyes, of such research would not be enough to explain that absence: the ancient lists preserve evidence of investigations no more "scientific"— for example, the collection of political constitutions gathered by Aristotle and his students. But one might call that an "external" explanation.

There is a deeper reason for the absence of an exhaustive census of animal species from Aristotle's biology. The nature of that biology is such that the very idea of a census for establishing universal correlations is alien to it, despite the unprecedented extent of the zoological information Aristotle reports. Aristotle's biology is usually teleological and/or apodictic. The correlations that the *Parts of Animals* tries to establish were not constructed inductively, since they are defended by appeal to the final cause and/or necessity. The correlations, Aristotle says, "conform to reason." [46] Thus, it is rational for all animals that have a bloody lung to have a bladder. In fact, "the abundance of blood is a sign of heat" (3.6.669b4), so much so that animals with bloody lungs "are the thirstiest animals, and need not only dry food but also much more humid food," and thus the necessity for a bladder (3.8.671a2). We can see immediately that for this sort of reasoning the examination of all cases is not methodologically necessary for the establishment of correlations: this is not an inductive reasoning that would be at the mercy of an unnoticed exception, but a rule based upon the "eulogia" of nature itself, and any exceptions are simply accidents in the order of the world. In an intellectual construction of this kind, the examination of particular cases does not really serve to establish the universal rules, but to illustrate them.

However, the universal correlations have some exceptions, either apparent or real. The exceptions that Aristotle counts as apparent are those reducible by appeal to analogy or to limiting cases. Thus, all blooded animals have four feet; but "instead of" the front feet, man has hands (4.10.687a7) and birds have wings (4.12.693b11)—that is the relationship of analogy. As for snakes, they do not have feet precisely in order to conform to the principle that blooded animals cannot have more than four points of movement. This is explained in the *Progression of Animals*, to which I refer here because it presents in its study of locomotion the same teleologico-etiological goal as the *Parts of Animals:*

> Evidence that snakes move like quadrupeds: in each part, the concave and convex alternate.   *(7.707b22)*

> The reason for the footlessness of snakes is that nature does nothing in vain but in every case looks out for the best possible arrangement for each thing, saving its special entity and essence; besides, as we have said before, no blooded animal can move at more than four points. Clearly, those blooded animals which are disproportionately long in relation to the rest of the nature of their body, as snakes, cannot have feet. For they are not the sort to have more than four (then they would be bloodless), but with two or four feet they would be practically immobile, so slow and useless the movement would have to be.   *(8.708a29, trans. Preus [1981])*

Between the fact of being blooded and that of having at most four limbs or motion-points, there is such a universal correlation that it can serve as the premise for a syllogism, as we see in the latter passage: every living thing that has more than four points of movement is bloodless; but serpents are blooded; therefore they have, even if that is not immediately obvious, four points of movement.[47]

Other exceptions, in contrast, are not so easily explained away when two correlations collide and contradict each other. As we saw above, it is reasonable for animals with a bloody lung to have a bladder. But it is no less rational, in Aristotle's view, that for animals with scales or shells the liquid residue in them is transformed into scale or shell: therefore, there is no need for a bladder to receive the liquid residue, and it would be in vain for nature to give them one. But turtles have a bladder: "In this case, nature has simply been stunted," Aristotle says.[48] Thus, he has to sacrifice one of his principles for the sake of the other:

> The reason for this anomaly is that sea turtles have a fleshy and bloody lung, like that of cattle, while land turtles have one that is disproportionately large.   *(3.8.671a16)*

In these latter cases, then, the liquid residue is too abundant to be entirely transformable into shell.

All this only confirms that if Aristotle is often a remarkable observer, he is not an experimenter in the modern sense of the word. On this point, Aristotle's frequent appeals to the facts — often while he accuses his predecessors of having ignored them — should not be misinterpreted. Robert Joly (1968) argues that Aristotle's biological "findings" are often "prescientific" in the Bachelardian sense of that word. And it is true that Aristotle is, in our modern eyes, singularly lacking in "axiological neutrality"; thus, Joly has no difficulty in showing that behind the error noted above concerning the sutures in the female skull lurks the male-centered *Weltanschauung* that Aristotle shared with his contemporaries. From this point of view, there is no room in Aristotle's biology for true experimental verification; the principles are not experimental, for the good reason that they are out of reach of an observation or an experience that might disprove or rectify them. I certainly agree with Cury (1960, 161) that "Aristotle's positive successes should not lead us to explain him in the name of some eternal and vague scientific spirit that he would have been the first to embody." Aristotle thought that he could make real progress toward what he called *episteme* by going from the question of how — or more exactly, from τò ὅτι ("the that") — to the question of why. The facts are therefore often submerged in an etiological discourse; the worst blindness is that of one who does not want to see. It was only by turning to observation and the collection of facts that the naturalists of the eighteenth century put their discipline on the road to science in the modern sense of the word.

Nevertheless, as I said above, there is a true Aristotelian zoology. Certainly, some overly zealous flatterers have erroneously thought that they could find in some of Aristotle's formulas a foretaste of experimental method; however, the very volume of his inquiries about animals shows that they are motivated by more than the simple wish to illustrate metaphysical speculation about nature. Joly rather oddly remarks (p. 252): "People often insist on the size of the biological inquiry. . . . I am quite persuaded, however, that some have exaggerated that size. Aristotle deals with about five hundred species of animals: that corresponds observationally with three one-thousandths of the species living in the geographical area directly accessible to him." I would simply like to note that Joly contradicts himself: he has well demarcated the difference between Aristotle's zoological research and modern natural science; but it is in relation to the eco-biological program of today's zoologists that Joly evaluates — and devaluates — Aristotle's results. As for comparing, as he does in the following paragraph in the same article, the species identified by Aristotle and the fifty or so species cited in the Hippocratic treatise *On Diet,* to leave the impression that Aristotle could very well have continued an ancient project of animal nomenclature, this looks like a bad joke. That is so even though that thesis was supported in detail by Carl Burckhardt (1904), some eighty years ago.

In fact, Aristotle does have a certain zoological encyclopedism. I have shown that the idea of an exhaustive census was foreign to his theoretical project, but there is little doubt that he and his students observed and described a great many species. This unusual curiosity is perhaps the origin of the opinion, based on shaky evidence despite Jaeger's attempt to stabilize it,[49] that Alexander, leading military invasions that were simultaneously investigational safaris, provided his old teacher with exotic animals. The fundamental theoretical reason for Aristotle's encyclopedism is his frequently reiterated thesis that reason alone cannot give an account of the real. On the one hand, logical consistency is not a sufficient guarantee of truth — madness, too, may be consistent.[50] On the other hand, because the sublunar

world is imperfect, and especially because of "the indetermi-
nateness of the matter and the plurality of principles" (*GA*
4.10.778a7), the real is always surprising, because it is not
always exactly what it should be. The most often cited example
(noted in the previous chapter) is that of sexual reproduction, in
which the product should always be a male resembling its father.

This necessity of empirical observation, which is in no case
experimental verification, appears clearly in the *Parts of Animals;*
not so much by the number of animal species mentioned (Aristotle names many fewer than in the *History of Animals*) as by
remarks that rely on a considerable set of observations about the
diverse forms and properties of each part. We find here again
that Aristotle tends to take a census — if not exhaustive, then as
large as possible — of animal families, which, as we have shown,
his teleological biology did not *in theory* require. There is more
in this encyclopedism than simply a desire to give illustrative
examples. Without ever having thought that he could claim to
have identified all existing animals, Aristotle does not seem to
doubt that his broad distinctions sufficiently combed through
the animal world so that no important variation in the nature of
one of the parts could have escaped him. It is thus that we should
understand a statement in the *Parts of Animals;* it is an incidental
comment, inadvertently tossed off, as it were, but it both poses
and decides a problem that seems theoretically fundamental to
us post-Linnaeans, for it says that the division of bloodless animals into molluscs, crustaceans, testaceans, and insects is exhaustive:

> Those that are called molluscs and crustaceans are very different
> from those [studied above]; for in the first place, they have incom-
> pletely developed viscera. Similarly, none of the other bloodless
> animals have developed viscera. There are only two other families
> of bloodless animals: testacea and the *genos* of insects.
>
> *(4.5.678a27)*

If, then, later zoologists have called themselves descendants of
Aristotle, it is especially because of his practice of observation —
an autonomous practice, in the sense that it is theoretically un-

motivated. We will now see that, in the *Parts of Animals,* the development of his research leads him, without any theoretical necessity, to distribute animals into distinct groups that modern zoology, from its own point of view, would often recognize as taxonomically valid.

The *Parts of Animals* in fact studies the variations of the different parts, and claims, wherever possible, to give the causes of the properties of the parts and of their variations. This brings us again to a fundamental realization: in the *Parts of Animals,* Aristotle often applies the scheme of division of *gene* into contrary *eide* as analyzed above; however, it is never in speaking of animals themselves, but of their parts, as is also the case in the passage from the *Politics* that we have already examined. Perhaps commentators have attempted to put Aristotle on the road to a taxonomy radically alien to him partly because they failed to notice this.

Certainly, as elsewhere, Aristotle uses the term *genos* to designate groups of animals that may be, from the point of view of modern systematicians, anything from a variety to a large class.[51] In contrast, he never says in the *Parts of Animals* (as he does say explicitly in the *Politics* passage) that there are several *gene* of digestive organs, etc., or that there is a *genos* blood, a *genos* mouth, a *genos* hair, etc., in which one must distinguish *eide.* However, Aristotle uses specific difference to divide each part, as previously defined by its function. He even gets to the point of applying the diairetic schema of *genos/eidos* to a function itself, as in the case of touch:

> For it seems to have in the highest degree [i.e., more than the other senses] numerous *gene,* and the perceptible that it perceives has numerous *contrarieties:* hot and cold, dry and moist, and other things of that sort. *(2.1.647a16)*

A few lines previously, Aristotle had written that "each perception *(aisthesis),* whatever it is, belongs to a unique *genos*" (647a6). Without entering into the details of a theory that appears in various versions in Aristotle's writings, we can say that here perception is a *genos* in that it is potentially perception of contraries. And, as usual, Aristotle divides at several levels: what he calls *genos* may be either a sense tied to a specific perceptual organ (647a7) or, within sense perceptions of the same kind (odor, touch, etc.), the various pairs of perceived contraries (hot/cold, etc.).

But Aristotle applies his diairetical method primarily to the *parts* in the *Parts of Animals*. And to mark this cardinal use of division in biology, he most often uses the term *diafora* or forms of the verb *diaferein:*[52]

> For the division *(diairesis)* of the homoiomerous parts shows a difference *(diafora):* in some cases the part is homonymous with the whole, and in some that is not true.   *(2.2.647b17)*

In this passage, division and specific difference are explicitly brought together in an expression that is, in any case, very difficult to translate.

As we should expect, following my previous analyses, the division of the parts (or functions) operates at several levels:

· *According to the more and the less:*

> In birds there is a difference between them in the excess or defect of their parts, that is, according to the more and the less.
> *(4.12.692b3)*

· *According to the "form";* so our passage about birds goes on thus:

> For some have long legs, others short; some have a wide tongue, others narrow; and the same for the other parts. But they differ only little among each other in the parts themselves; in contrast, they differ from other animals by the form of the parts.   *(692b5)*

· *The difference can be analogical:* there are many examples in which Aristotle asserts that some animal family has the analogue of an organ that some other family possesses.

But this division of the parts according to specific difference is not only valid at all levels, it is also multidimensional, in that, applied to the same part, a division makes separations among animals and also within the same animal:

The differences that distinguish these parts from each other are for the sake of the better. The comparison of blood to blood is a good example: one is lighter, another thicker; one is purer, another muddier; and one is cooler, another warmer, even in the parts of the same animal—for the blood in the upper parts differs from that in the lower parts in respect of these differences—and from one animal to another. *(2.2.647b29)*

Finally, most often, the differences between the parts should be put together with the general differences of the animals, according to the classes to which they belong:

In some the liver is separated into several lobes, in others it is, rather, undivided; this difference is found even among the viviparous blooded animals, but it is even greater between those and the fish and oviparous quadrupeds, which also differ among each other. *(3.12.673b17)*

The central project of Aristotelian biology is thus what we could call an "etiological moriology," and it is in the *Parts of Animals* that its "scientific" construction is carried out. We have already seen, at the beginning of the present chapter, that the diversity of animals should be considered secondary to the diversity of their parts, which is more radical (Aristotle would say it is "prior"). Thus, it is not very surprising that from the Aristotelian point of view the process of division applies fundamentally to the parts. Just as animals differ basically because of the differences between their parts, the diairetical schema applies to animals themselves only derivatively—that is, only because it is applied, fundamentally, to their parts. That realization may lead us to reconsider many passages. Let me give just one example, intentionally a debatable one, in that the traditional interpretation is neither absurd nor grammatically impossible; it is taken from the first (methodological) book of the *Parts of Animals*. Having reminded us that biological research is formal and final, Aristotle writes:

'Αναγκαῖον δὲ πρῶτον τὰ συμβεβηκότα διελεῖν περὶ ἕκαστον γένος. ὅσα καθ' αὑτὰ πᾶσιν ὑπάρχει τοῖς ζῴοις, μετὰ δὲ ταῦτα τὰς αἰτίας αὐτῶν πειρᾶσθαι διελεῖν. *(1.5.645b1)*

A. L. Peck translates this as follows:

> First of all, our business must be to describe the attributes found in each group *(genos);* I mean those "essential" attributes which belong to all the animals, and after that to endeavour to describe the causes of them.   *(1.5.645b1)*

I prefer to assume that the term *genos* here designates the different sorts of parts, and not the different groups of animals. For it is not likely that Aristotle would propose to divide (that is the sense of *dielein,* and its translation by "describe" masks the problem without resolving it) the attributes common to all animals according to genera (or species) of animals; it is more likely that he intended to divide the common attributes by distinguishing them into *gene,* which is confirmed by the rest of the passage ("But many parts are common to many animals") and by the very procedure adopted in the *Parts of Animals.* Thus, I would translate the passage as follows:

> It is necessary, first, to differentiate the attributes that belong essentially to all animals by differentiating each *genos* of them, and after that we must try to differentiate the causes of these attributes.

Nevertheless, although it claims to depend primarily on metaphysical principles, this study of the parts ("moriology"), which is constructed principally by finding specific differences within the various *gene* of organs and functions, presupposes an ordering of animals. That ordering — empirical and not "scientific," as we have seen — is the background of Aristotle's text, though we are not told how it was constructed. Only rational zoology, the only real zoology (for Aristotle, too, everything that is rational is real), can construct *a priori,* by the combination of the various species of each part, the entire animal world. It is this ideal zoology that appears, as a project, in the passage from the *Politics* that we discussed above; referring to biology only for comparison, Aristotle is not there tangling with the ultimate unpredictables of the real. Because of the ineradicable margin of contingency that he meets in *physis,* the naturalist should also have recourse to empirical investigations. Ordering animals is one such investigation that, one might say, slips through the cracks of an impossible *a priori* zoology. Finally, we should notice that this limitation, this gap in relation to the ideal zoology, is not attributed by Aristotle to the weakness of the available means of investigation, whether material or intellectual, but, as we have seen, to the ontological deficiency of the sublunary world.

Finally, the last stage of Aristotelian zoology, the treatise on the *Generation of Animals,* studies the moving cause, as the programmatic passages quoted above announced, especially that at the beginning of the *Parts of Animals* (1.1.640a15). It is, by the way, interesting to put this treatise beside Books 5 to 7 of the *History of Animals,* where the reproduction of living things is studied. One again finds the same relation as between the *History* and the *Parts of Animals* concerning the other functions: the relation of description to causal explanation. It is one thing to describe the genital organs, the ways of coupling, and so on, and another to tell why generation, its differences, its successes, and its failures exist. Again, I find a profound incomprehension of this complementarity by Pouchet, when he writes: "From the first lines of the treatise *On Generation* (1.1), the author declares that he has never written about this material, and yet at every moment we find references to the *History of Animals.* We rely on this proof, among others, of the disorder of the Aristotelian collection." [53]

We know, finally, that the fifth and last book of the *Generation of Animals* studies "the characters by which the parts of animals differ" (5.1.778a16). These characters are not determined by the final cause, but "have their cause in the movement and in generation" (1.778b14) — that is, in the activities that are exercised upon the embryo at the moment of its formation. Thus, there is "the blue or dark color of eyes, sharp or deep voice, differences in color of the hair or feathers" (1.778a18). This is a causality that I would call non-teleological.

There is no doubt that this last study was meant to conserve and even to reinforce teleological biology by plugging up one of its important holes. This theory of the "characters" nevertheless reinforces the tendency toward observation, whose relative but real autonomy, as we have seen, makes of Aristotle, in our modern eyes, a biologist in the modern sense of the word. But this study of non-teleological characters is carried out within the same perspective and with the same means as the rest of Aristotle's biological studies. For example (an important fact for us), the parts are taken here, too, as *gene* and divided according to the habitual schema:

> The eye is for something, but it is not blue for something, unless that is a characteristic peculiar to its *genos*.   *(1.778a32)*

In Aristotle's etiological and apodictic zoology, animal classifications are thus outside science. They are affected by the same ambiguities as other empirical procedures. I cannot agree with Lloyd, finally, when he says that, for Aristotle, "zoological taxonomy becomes a *problem,* with the possibility of further critical discussion — the evaluation of the grounds for theories and beliefs — and of research" (1983, 205). Those studies — unworthy of *episteme* — nevertheless gained for Aristotle "marvelous pleasures," as he himself confesses (*PA* 1.5.645a9). But they are necessitated more by the deficiency of the world than by its perfection. These empirical procedures, despite their theoretical poverty from the Aristotelian point of view, were what modern naturalists would praise the most in his biological work.

CONCLUSION

T<small>HUS, ARISTOTLE CLASSIFIES</small> animals and he classifies them well — much better, at any rate, than most of his successors. Every modern man, aided by the acquisition of both taxonomy and comparative anatomy, grasps without difficulty the principal reason for the "objectivity" of Aristotle's classifications: by getting down to the parts and functions, he escaped the seductions of the immediacy of animal characteristics and behaviors. But no one should ascribe to him any taxonomic project, even an unformulated one. Aristotle and the systematicians of the classical period might appear to have set themselves related questions: through the notion of difference, they try to see how the Same and the Other are combined in nature. Indeed, it does seem that there is a question, hardly formulated, underlying Aristotle's biological research: that of the unity within phenomenal diversity of living forms, "phenomenal" in both the technical and the ordinary senses of the word. I have noted that at the level of each living thing, Aristotle's general explanatory principle is that "nature does nothing in vain." But what is the principle at the level of all living things taken together? At least one passage, in the *Politics,* sketches a teleological explanation of that diversity, but it is an attempt that remains limited, and Aristotle himself notes that it is partial:

> Plants exist for the sake of animals, and the other animals exist for the sake of man: domestic animals for his use and food; wild animals — if not all, then most of them — for his food and his other needs, such as clothing and tools.   *(1.8.1256b16)*

But the idea that diversity is itself a perfection, sometimes attributed to Aristotle, is doubtless a later development, Stoic perhaps, Thomist surely. Thus, we read in the *Summa Contra Gentiles* (2.45.1): "But created things cannot achieve the perfect likeness of God according to a single species of creature: because, since cause exceeds effect, that which is simple and unified in the cause is composite and multiple in the effect." We must, however, recognize that Thomas Aquinas, a careful interpreter of Aristotle, does not attribute this doctrine to him directly.

In fact, we should not say that Aristotle distinguished an *order* in nature, in the sense in which taxonomists would postulate such an order. For the systematicians, in fact, the concept of order is the fundamental *response* to questions concerning the living world, in that this order is the guarantee of the intelligibility of the world. Michel Foucault (1966, 71) has given a good account of that period, limited in time, which saw the rise and fall of "taxonomy":

> The fundamental element of the classical *episteme* is neither the success nor the failure of mechanism, nor the correctness or impossibility of mathematizing nature, but rather a link with the *mathesis* that remained, until the end of the eighteenth century, constant and unaltered. This link has two essential characteristics. The first is that relations between beings are indeed to be conceived in the form of order and of measurement, but with this fundamental imbalance, that it is always possible to add to the problems of measurement the problems of order. . . . And the Leibnizian project of establishing a mathematics of qualitative orders is found at the very heart of classical thought. . . . But, on the other hand, this relationship to *mathesis* as a general science of order does not signify that knowledge is absorbed into mathematics . . . ; on the contrary, . . . one

sees a certain number of empirical domains appearing that, until the present, had not yet been either formed or defined. . . . Thus appeared general grammar, natural history, the analysis of wealth.

(Translation adapted from 1973, 57)

For Aristotle, on the contrary, this ordered diversity of living things is a problem. As I have said, this ordered diversity does not manifest in itself the perfection of nature; rather, Greek thought had the tendency to see diversity as a mark of imperfection. And in the eyes of Aristotle and his contemporaries, one of the metaphysical advantages of a "moriology" was precisely that it reduced that diversity by showing that in the final analysis, if one goes beyond the immediate appearances, it is the Same that wins.

One may wonder, then, why Cuvier was so enthusiastic about Aristotle. We saw at the beginning that Cuvier claimed descent from Aristotle rather than from his immediate predecessors, the eighteenth-century taxonomists. In a general way, the most obvious relationship between Aristotle and Cuvier comes from the fact that, in contrast to the taxonomical naturalists, they are both biologists. As Foucault says:

> What to Classical eyes were merely differences juxtaposed with identities must now be ordered and conceived on the basis of a functional homogeneity which is their hidden foundation. When the Same and the Other both belong to a single space, there is *natural history;* something like *biology* becomes possible when this unity of level begins to break up, and when differences stand out against the background of an identity that is deeper and, as it were, more serious than that unity.[1]

This deeper identity is, for Cuvier's comparative anatomy, that of organs and functions. In the words of Joseph Chaine (1925, 270):

> For Cuvier, . . . the real elements of comparison in anatomy are the organs. Thus, he gives his entire attention to them and, taking them individually, compares them to themselves in all the modifications that they manifest in going from one species to another. But at the same time, as he was convinced that every organ is characterized

> by the function it fulfills, he thought that the rigorous determina-
> tion of function was a capital fact. . . . He proposed that every
> organism is an harmonious combination of organs, but that this
> harmony is nothing but a consequence of that of the functions.

The relationship between Cuvier and Aristotle may be clarified
by noting (1) that for Cuvier "the functions that compose the
economy of the animal can be related to three orders,"[2] which
are those of "animal functions" of perception and movement,
"vital functions" "that serve to nourish the body" (p. 17), and
the function of generation; and (2) that across the modifications
of organs from one species to another, "not all the organs follow
the same order of degradation; one is most perfect in one species,
and another is perfect in quite a different species; so that if one
wished to order species according to each organ considered indi-
vidually, . . . it would be necessary to calculate the resultant
effect of each combination [of organs]" (p. 60). When we read a
passage like this, we are struck by the similarity between the
views of Cuvier and those of Aristotle, particularly in respect to
the *a priori* zoology outlined in the passage of the *Politics* that we
have discussed.

As for the problem that concerns us, the classification of
animals, Aristotle and Cuvier have many points in common.
The fifth and last article of the first of the *Lectures on Comparative
Anatomy* is entitled "Division of the Animals According to the
Whole of Their Organization." Cuvier begins by giving a justification of his classificatory enterprise that looks very Aristotelian. He wants to avoid repetitions in the study of the organs;
then, having indicated that there are two competing methods of
classification, he writes (p. 68):

> The formation of the methods is the objective of natural history,
> properly so-called; anatomy receives them, so to speak, ready-made;
> it is from natural history that anatomy gets its first directions; but
> anatomy does not delay in returning the lamp whence it has re-
> ceived it; it is even the strongest test of its goodness.

Historically later than natural history, comparative anatomy,
the study of organic and functional differences, is in fact its

logical foundation. In contrast to natural history, comparative anatomy goes beyond representation to "explain the nature and properties of each animal" (p. 37). Here, too, we find a procedure close to Aristotle's etiological study.

Pedagogically, then, for someone who undertakes the comparative study of living things, comparative anatomy depends upon natural history. But *theoretically,* natural history depends on comparative anatomy. On this topic, recall Aristotle's distinction between that which is better known *per se* and that which is better known by us. So by giving natural history a purely instrumental role, Cuvier eliminates the possibility that it could have the goal of developing a "natural method." As in the case of Aristotle's classifications of animals, this is a theoretical devaluation. Although Linnaeus handed on to subsequent generations the task of completing his taxonomical method, at least he was sure that such a method existed, and that the genera and species, completely escaping the arbitrariness of human thought, were the work of nature itself.[3] Here again, Cuvier does not follow Linnaeus: he understood the necessarily empirical and incomplete character of every classification; and like Aristotle in this respect, it is without any methodological justification that he posits that all animals are divided into four branches: vertebrates, molluscs, arthropods, and radiates. Thus, classification had ceased being a theoretical task to become again, as it had been for Aristotle, a tool for research *at another level.* Darwin would soon come along to give the concept of species a purely conventional character. Taxonomy would then be dead.[4]

Cuvier's taking up of an Aristotelian project should not, however, delude us. The anti-taxonomic agreement between our two authors should not allow us to forget that there are such great differences between them that they are historically incommensurable. Aristotle and Cuvier are found on opposite sides of "taxonomy," which constituted a deep rift in the history of the life sciences.

The naturalists of the classical epoch had to overturn Aristotle's teleological "moriology" so that "Taxonomy" could come into being. As always in the history of science, the critique that preceded progress began from a lack of understanding. By the seventeenth and eighteenth centuries, no one had any comprehension of what Aristotle meant by "final cause." Those who undertook to criticize teleology, like Galileo and Spinoza, thought of it either as circular or as a *post hoc* justification; others, like Bernadin de Saint-Pierre, thought that they could defend it with poetic naiveté; neither understood Aristotle's concept of teleology. To imagine that Cuvier's comparative anatomy — in a sense, the negation of a negation of Aristotelianism — is in perfect theoretical agreement with Aristotle's zoology would be an anachronistic, even atemporal, reading. Cuvier's critique of taxonomy surpasses all previous biological theories: we should not confuse "never again" with "not yet." For Cuvier, the theoretical goal of the taxonomists was no longer satisfactory because he sought the organic structure beneath the apparent characteristics; nevertheless, taxonomy remained an historical and methodological premise of his comparative anatomy.

Neither a taxonomist nor a precursor of comparative anatomy, Aristotle constructs an apodictic and teleological biology based upon his metaphysical principles. Continuist readings of his work, those that make a precursor of him, can be explained by many factors. Let me mention two: one depends upon us, the other may be traced to Aristotle. In the first place, in a very general way, the attitude of modern European culture toward Greek and Roman antiquity can only obscure the true nature of theoretical affiliations between these two worlds. Perhaps only now, when "classical studies" have gone thoroughly out of fashion, can we, paradoxically, benefit by no longer thinking of the Greeks as our fathers. However, a first and decisive break was already made by people like Louis Gernet and his students, or Eric Robertson Dodds, when they applied the concepts of social anthropology to ancient Greek society, thus ending the ethnocentrism that had been so costly for historians. Filial piety is an insurmountable epistemological obstacle.

Another obstacle to the correct appreciation of the relations between Aristotelian biology and later natural sciences comes from the tension between the logical demands of Aristotle's metaphysical approach to nature and his observational habit, which one might even call pre-experimental. Aristotle's observations are so numerous that they tend to take on a certain autonomy in relation to the *a priori* principles that they were meant to illustrate. Nevertheless, we must firmly resist the temptation to make Aristotle a naturalist in the modern sense. Behind the apparent similarity in their language, there lies an insurmountable gap between Aristotle and Cuvier that separates their goals and prevents them from communicating. Let us look briefly at that gap, as it concerns the principles of biological research. Cuvier writes (p. 10):

> The goodness with which nature has treated all her productions does not permit us to believe that she deprived beings capable of perception — that is, of pleasure and pain — from being able to flee the one and pursue the other up to a point.

This passage seems perfectly Aristotelian. We think we recognize especially the teleological principle according to which nature always accomplishes the best. The last words, which seem to mark the resistances that such a principle ought to conquer in order to actualize itself effectively, also sound Aristotelian. But the similarity is illusory. For alongside this principle (transcendental in the Kantian sense), we also find in Cuvier a scientific practice that obeys only its own rules. So much so that whereas in Aristotle observation was, as it were, blinded by the tyrannical prestige of the principle, in Cuvier the principle of the goodness of nature does not direct and does not mislead observation. Aristotle, in contrast, generally keeps only those facts that conform to his principles and tends to ignore or minimize the others.

Does not the objectivity of the distinctions that Aristotle drew in the animal world, praised by Cuvier (they "left very little to be done in later centuries"), come principally from the "nonscientific" status of Aristotle's animal classifications? Because Aristotle found them too negligible, from a theoretical point of view, to base them in rationality, because he has not even shown us how he arrived at these classifications, they gain a certain independence from the metaphysical and ideological principles that otherwise regulate his biology. Thus, we have in the animal classifications the product of Aristotle's intellectual work in the purest possible form of empirical observation free from all metaphysical meddling. It is for that reason that these classifications would later be able to offer, independently of Aristotelianism, a basis for later naturalistic study. These animal classifications, with so little importance in Aristotle's own eyes that he never tells us how he constructed them, thus became the most precious legacy handed down to later zoologists.

NOTES

INTRODUCTION

1. Louis (1975), p. 29. This work appeared two years after the last volume of the biological treatises edited and translated by Louis in the Budé series.

2. Meyer (1855). See chap. 11, "Principien der Entheilung." The work of Meyer remains one of the best of those on the whole scope of Aristotle's biology.

3. A "natural" method of classification was first constructed in botany: "For more than two centuries, the classification of plants was better than that of animals," writes Henri Daudin (1926, 21). But in botany, as Émile Guyenot points out, the only method that can be called natural is that which puts into relation not only one part of a species with that of another (the parts for fruiting, for Linnaeus), but all the parts. Thus, Guyenot quite properly writes (1941, 29): "Although he has been systematically ignored and shamelessly plagiarized, the real founder of the natural method was the inspired French botanist Michel Adanson."

4. See, for example, the term ὀνοματοποιεῖν with the references given by Bonitz (1870) 515b29 et seq. Note that in *Cat.* 7.7b12, an example taken from zoology (the wing) leads Aristotle to envisage the creation of new terms; and that in *Top.* 8.2.157a29, this recommendation has an incontestable taxonomic value, since it is a matter of constructing adequate universal groups by induction (τῇ δ' ἐπαγωγῇ, 157a19), and it is in order to designate these universal classes that one must not hesitate to invent new terms. Plato had already demanded for the philosopher this liberty of verbal invention. Thus, in the *Sophist* he justifies the use of the term δοξομιμητική because "there is a serious shortage of names" (267d, trans. Cornford).

5. Manquat (1932, 104): "But Aristotle uses only the vulgar names of the animals he cites. Even so, these names seem to us perfectly

scientific, because Aristotle spoke Greek, and our scientific terminology is founded on Greek; let us not forget that those words represented for him and his contemporaries popular forms."

6. See *HA* 1.1.487a32: καλῶ δ' ἔντομα ὅσα ἔχει κατὰ τὸ σῶμα ἐντομάς ("I call 'segmented' the animals whose bodies have segments"). Examples of this kind are in fact rather numerous. Thus, the term λόφουρα designates a class that groups horse, ass, and mule, and that would therefore be close to what we call equids: it is a descriptive term — "animal with a mane" (of horsehair, if you like), as the examples make clear — hardly used except by Aristotle and Theophrastus (*HP* 2.7.4; 3.10.2; 5.7.6). As for the word στεγανόποδα, it may have been invented by Aristotle himself (literally, "with feet that let nothing pass"), which he opposes to σχιζόποδες ("with split feet"). See, for example, *HA* 8.3.593b15, where it corresponds linguistically to our "palmate footed." Aelian uses it once (*Nat. An.* 11.37). Sometimes Aristotle notes that a group lacks a name and he suggests one, in Greek, but he does not normally continue to use it in the biological works; thus, in *HA* 9.40.623b5: ἔστι δέ τι γένος τῶν ἐντόμων, ὃ ἑνὶ μὲν ὀνόματι ἀνώνυμόν ἐστι, ἔχει δὲ πάντα τὴν μορφὴν συγγενικήν. ἔστι δὲ ταῦτα ὅσα κηριοποιά, οἷον μέλιτται καὶ τὰ παραπλήσια τὴν μορφήν ("There is also a group of insects, whose members do not have a unique name, but which all have a related form; these are the 'comb-builders,' as the bees and other insects with a related form"). Janine Bertier (1977, 48) notes that the term ὀστρακόδερμα does not appear earlier than the *Historia Animalium*. As for the selachians, note that the term can be found in the form ἰχθύσι σελάχεσι in the second book of the Hippocratic *Illnesses* (no. 50).

If the groups named above appear, after all, to have an objective coherence comparable to that of groups named nowadays, there also exist in Aristotle's writings groupings that one could call more "phenomenological." Thus, when Aristotle speaks of "ruminants," τά ζῷα μηρυκάζοντα (*HA* 9.50.632a33), it is purely and simply a description of behavior, since he includes fish in it (632b9).

Finally, there are numerous groups that Aristotle designates by a word derived from the name of the species most representative of the group; for example, κορακώδης or κορακοειδής, which corresponds to the English "corvine."

7. ἄναιμος ("bloodless") is used by Plato to describe the hoof (*Protag.* 321b) and the lung (*Tm.* 70c). τὰ ἔναιμα is found in the *Timaeus* (81a6); Cornford translates this as "the substances in the blood." In Herodotus (3.29) the word is used in the sense of "made of blood." Cambyses cries out, while slaughtering the Apis bull: τοιοῦτοι θεοὶ γίνονται ἔναιμοί τε καὶ σαρκώδεες ("Such gods are made of flesh and

blood"). One also finds the term in the Hippocratic corpus in the sense of "full of blood" in designating an organ (*VM* 22) or "bloody" in speaking of wounds (*Fract.* 24). The *On Diet* uses the word ἄναιμος four times, once in the comparative (2.46.2, 3; 49.2, 3) and once in the superlative ἐναιμότατα (2.49.3) to designate the more or less bloody complexion of an animal's flesh. Notice that we are here dealing with three treatises that are thought to be "old," possibly even preceding Aristotle. In a treatise unanimously thought to be Cnidian, *Illnesses* 2, the author uses in two parallel texts (chaps. 4 and 7, Littré vol. 7, pp. 10 and 30), for designating the blood vessels that surround the brain, the expressions περὶ τὸν ἐγκέφαλον φλέβια and τὰ φλέβια τὰ ἔναιμα τὰ περὶ τὸν ἐγκέφαλον. Some manuscripts have ἐν αἵματα instead of ἔναιμα. Theophrastus speaks of φάρμακον ἔναιμον (*HP* 4.7.2), "sistendo sanguini medicamentum" (trans. Friedrich Wimmer), "a drug for stanching blood" (trans. Sir Arthur Hort), but one finds at least three times in Theophrastus's text the terms in question used in their Aristotelian sense of "blooded animals" and "bloodless animals": *HP* 1.2.5; *Sens.* 3.23; and frag. 171.6. It is interesting to notice that we find Aristotle himself using the terms *enaimos* and *anaimos* in the "pre-technical" sense, i.e., in a "pre-Aristotelian" way, of "provided with" or "lacking blood" (see *GA* 1.6.718a12).

8. It was only in the eighteenth century that naturalists perceived clearly that the taxonomic project could be carried out only at the cost of recoining names. François Dagognet has correctly pointed this out in his work on the history of taxonomy (1970, 29): "[In botany,] the invented name is really not so much a label as an advance indication of the characteristics and properties of the species. It makes possible not so much the recognition as the precognition of the plant." (The French also has a play on *énoncer* ["label"] and *annoncer* ["advance indication"] in the first sentence.) In zoology, writes Dagognet (1970, 118): "One hopes to imprison the beast in a clear syllable, inside a single sound. An invented name signifies the surest possession, a kind of ultimate thesaurization. In its way, it imposes a matrix and sorts out essential from accidental, truth from that which hides the truth." And Dagognet is surely right when he says (1970, 35) that when Tournefort rejected invented names and "hypocritically preached obedience to custom," this rejection was simultaneously the cause and the consequence of his failure in classification.

9. Probably taxonomy will not really be completed until it is understood that its project of a unique natural nomenclature is impractical.

10. Cuvier (1841). The lectures on Aristotle are the seventh and eighth in the first volume. Most commentators quote together the

judgments of Cuvier and Darwin on Aristotle, as two comparable notices, equally enthusiastic, of an intellectual debt to the founder of the Lyceum. They rely on a letter of Darwin to William Ogle, translator of Aristotle's *Parts of Animals*, dated February 22, 1882. Generally they cite only the famous line, "Linnaeus and Cuvier have been my two gods, though in very different ways, but they were mere schoolboys to old Aristotle." (Charles Darwin, *Life and Letters*, vol. 3, p. 252; cited from Ross [1949, 112].) In a recent note, Simon Byl (1973) has put things into perspective by showing that Darwin wrote this letter to thank Ogle for sending him a copy of the translation, and that Darwin expresses not only his inevitable admiration but also, and perhaps even more, his astonishment at Aristotle's errors; in fact, he admits to having read not more than a quarter of the translation that Ogle had sent him. Certainly, Darwin admired Aristotle, and it would have been difficult for him to do otherwise, but one cannot find in him both the knowledge of the texts and the awareness of theoretical affiliation that one finds in Cuvier.

11. Perhaps modern interpreters have been a bit too quick to imagine that the division between blooded and bloodless is only another way of stating our present division between vertebrate and invertebrate. There are two different points of view here, and they do not give the same place to various functions in the animal economy: the one fundamentally defines the animal by its nutritive function, the other by the presence or absence of a bony armature. Furthermore, it was Aristotle who constructed the concept of "vertebrate": "The vertebral column (ῥάχις) is the principle of all animals that have bones" (*PA* 2.9.654b12). Here again, Aristotle chose a point of view that is not ours.

12. τὸ τῶν βατράχων γένος, *PN* 470b17, can perhaps be translated "the race of frogs" ("la race des grenouilles"—Mugnier), but in no case "the class of batrachians" ("la classe des batraciens"—Tricot).

1. DIVISION AND DEFINITION IN ARISTOTLE: THE MEANING AND THE LIMITS OF THE CRITIQUE OF PLATONISM

1. All the studies devoted to zoology before Aristotle show, often more than they would like, how much we are reduced to conjectures about those texts. See, for example, on these pre-Socratics, Leboucq (1946) or Kucharski (1964).

2. The most interesting of these external bits of evidence is the fragment of the comic poet Epicrates, cited by Athenaeus (2.59d), in which one sees young students of the Academy, in the presence of Plato

himself, exercising themselves at defining animals and plants by the method of division, and trying to characterize the pumpkin despite the mockery of a Sicilian medical man, a skeptical materialist who "farted in their face" (κατέπερδ' αὐτῶν)(59f). Strycker (1932) was certainly right to point out, against the interpretation of Paul Shorey, the preponderant place of the method of division in the later dialogues of Plato. The critique is to be found in the second part of his article, pp. 230ff.

3. Did Aristotle at some time compose a work consecrated to Platonic divisions? One gets that impression from a certain tradition, issuing notably from Diogenes Laertius, who includes a *Divisions* in his list of Aristotle's works. We should certainly adopt the prudent attitude of Moraux (1951, 85), who recognizes that "one cannot hope to find in it either a faithful exposition of Plato's researches or a work of the most talented of his disciples."

4. I shall argue in a later chapter that Aristotelian biology is *apodictic,* and we shall see particularly that zoological observation should end in the presentation of universal correlations that can serve as premises. That is confirmed by the frequency of biological examples in Aristotle's "logical" writings. See, e.g., *APo.* 2.17.99b5: "Every animal without bile is long-lived, but four-footed animals do not have bile."

5. *PA* 1.3.643b10, trans. Peck. Identical references to "common sense" are found at 1.4.644a16 and 644b2. Cf. *Mete.* 1.13.349a27.

6. Louis (1955a, 299): The first book of the *Parts of Animals* "has been studied with great care and penetration by J.-M. LeBlond, who, however, has insisted too much on the negative aspect of the considerations that Aristotle develops. Without doubt, Aristotle essentially aims at accenting the defects of the methods of classification used by the scholars of his time, and in particular those of the dichotomic method."

7. The denunciation of the repetitive character is found at 642b9 and is repeated at 643b36. The splitting up of genera is criticized at 642b18, 643b3, and 643b14, but one may combine this criticism with the one that accuses dichotomy of not dividing according to substance but according to accidents, even if these are *per se* accidents (642a27). For division according to privation, which leads dichotomy to distinguish species in non-being, see 642b23. The finest and most complete analysis of the contents of Aristotle's critique of dichotomy is to be found in Balme (1972, 101–119); I have nothing further to add to it on this subject.

8. The text of Chapters 2 to 4 has lacunae or conjectures in at least five important places: 642b35, 643a9, 643a24, 643b3, and 643b9.

9. Pierre Aubenque (1962, 113) may be right in supposing that Aristotle's confidence in traditional distinctions was based upon his

confidence in the fundamental rightness of ordinary language: "Relying on words, one may be confident of not completely missing the truth of things; the very fact that human beings use them, and use them to good purpose, proves by itself that words adequately fulfill their function of designating. That explains the confidence that Aristotle, as scientist, seems to have in the classifications provided by popular language; the success of a common designation indicates that these designations are not arbitrary and that the oneness of the name should correspond to the unity of a species or a genus."

10. Gilson (1971, 14) has well described the philosophical import of this distinction, which, in its way, circumscribes the biological field: "Why is there heterogeneity in the structure of certain beings? Because they are alive, . . . the heterogeneity of the parts is requisite for making possible the causal action upon oneself that characterizes the coming-into-being of living things."

11. Plato several times explicitly indicates this binary character of his method of division: e.g., in the *Sophist* 264d11, σχίζοντες διχῇ τὸ προτεὺὲν γένος; and in the *Statesman* 262e4. Aristotle, by contrast, seems to remain faithful to the ordinary sense of διαιρεῖν in often dividing into more than two: e.g., in the *Politics* he divides goods into external goods, goods of the body, and goods of the soul (7.1.1323a24), or melodies into ethical melodies, active melodies, and enthusiastic melodies (8.7.1341b32). But this practice can also be found in the "logical" and biological works: thus, in *Categories* 13.14b37, Aristotle divides animals into winged, footed, and aquatic; and I shall analyze more closely below *Parts of Animals* 1.3.643b12, in which Aristotle accuses dichotomy of being incapable of employing several differences simultaneously. In fact, we shall see in the next chapter how Aristotle's own method of division is both binary and multidimensional.

Besides the fact that Plato recognizes that division into two is not always possible (see *Statesman* 287b10), was a non-dichotomous diairesis practiced in the Academy itself? The division of goods into three (psychic, bodily, external), which we have noticed in Aristotle, is also to be found in the *Laws* (3.697a6: τριχῇ διελεῖν). Furthermore, Diogenes Laertius gives a long list of divisions that, *according to Aristotle* (φησιν ὁ ᾿Αριστοτέλης, 3.80), Plato made: there are three sorts of goods, three sorts of friendships, five kinds of constitutions. . . . Diogenes Laertius indicates these divisions sometimes by the verb διαιρεῖν (e.g., 3.88: ἡ μουσικὴ εἰς τρία διαιρεῖται), but most often he says that there are so many εἴδη of something.

12. 488a7: ὁ δ᾿ἄνὺρωπος ἐπαμφοτερίζει. I shall discuss the term ἐπαμφοτερίζειν at a later point. It is difficult to ascertain to which solitary men Aristotle is alluding here, when we know that he declares

in a famous passage of the *Politics* (1.2.1253a29) that the man who would be able to live alone would be "either a monster or a god." "Wild men" are discussed in the following note.

13. *HA* 1.1.488a26 says: "Further, we find some are tame, others wild; some are always tame, e.g., man and the mule; some always wild, as the leopard and the wolf; some can also be quickly tamed, e.g., the elephant. Here is another point of view (ἔτι ἄλλον τρόπον) [Peck omits this phrase]: for any tame kind of animal exists also in the wild state, e.g., horses, oxen, swine, men, sheep, goats, dogs" (trans. Peck, with amendment noted). Compare *PA* 1.3.643b4, which says: "Practically all (πάντα γὰρ ὡς εἰπεῖν) the tame animals are also found as wild ones: e.g., man, the horse, the ox, the dog (in India), swine, the goat, the sheep."

In the first place, there seems to be a contradiction in the first passage, since Aristotle says in it both that man is always domesticated and that he can be found in a wild state. Most editors have believed that they could avoid the problem by modifying the text: in the place of ἄνϑρωπος Louis writes ὄνος, Dittmeyer γίννος, D'Arcy W. Thompson ὄνος ἵππος; Peck puts the entire phrase with the first mention of man in brackets. Most editors accept the second mention, although some purely and simply omit it, as did Guillaume de Moerbeke. In my opinion, Aristotle propounds in this passage two different opinions that he does not choose between, because that is not the issue; he says, in any case, "here is another point of view." Lacking information about Aristotle's position, we see from this passage that at least the question of wild men was debatable in his day. It is possible that wild men furnished the Greek imagination with an inverted image even more fertile than that offered, according to François Hartog (1979), by the Scythian nomads. Perhaps one may find traces of this debate in the beginning of Plato's *Sophist* (222b).

At any rate, the text of *HA* tends to support my thesis: it criticizes Platonic dichotomy by showing that it wrongly divides into domesticated and wild.

14. François Nuyens (1948, 198–201) thinks that *PA* 1 was written after the three other books, which he takes to be later than the *HA* (p. 158).

15. *HA* 8.2.589a13. "Define further" translates προσδιοριστέον (589b13). This is a rather complicated passage in which several divisions are superimposed: animals-with-lungs/animals-with-gills; blooded/bloodless; animals-feeding-in-the-water/animals-feeding-on-land. In fact, there is a combination here that could allow a classification of aquatic animals. Thus, the triton or water-newt (κορδύλος) is the only animal with gills that goes looking for its food on land.

16. Aristotle uses the word διχοτομία (or forms of the verb διχοτομεῖν) five times in PA 1: 642b22, 643b13, 643b25, 644a9, and 644b19. It is a rare word that is not often found in Aristotle outside this passage: The Bonitz *Index* notices five other uses of the word, if uses repeated a few lines apart count as one. Outside Aristotle the word is rare except in later scientific or philosophical texts. The examples of the use of διχοτομία given by LSJ come from Geminos, Simplicius *(Commentary on Aristotle's "Physics")*, Theon of Smyrna, and Iamblichus. In contrast, Plato never uses διχοτομία, and the verb διχοτομεῖν is found only once *(Statesman* 302e7). Aristotle, having left Platonic orthodoxy, speaks of the Academy in non-Academic language. Sometimes he does use the word διαίρεσις but specifies it with expressions such as εἰς δύο διαφοράς (642b6), εἰς δύο (642b17, 643a17), εἰς τὰ ἀντικείμενα (642b35, but here the text is perhaps corrupt), τοῖς ἀντικειμένοις (643a31), δίχα (642b27, 644a11), and διαφορᾷ μιᾷ (643b9).

17. Pellegrin (1981). For a more complete study of Aristotle's division, see Artur von Fragstein (1967).

18. ὁποιαοῦν διαφορᾷ μιᾷ. Louis's correction ὁποιονοῦν (643b9) seems both less acceptable from the point of view of the sense and farther from the manuscript readings, which give ὁποιαιουν, ὁποιανοῦν, ὁποιαοῦν. Most other editors write ὁποιαοῦν, which they then refer to διαφορᾷ μιᾷ. That correction is acceptable.

19. Michael of Ephesus uses a nice expression to tell us that Aristotle does not accept common opinion without discussion: σύνηϑες γὰρ αὐτῷ μαρτύρασϑαι καὶ τὴν τῶν πολλῶν ὑπόληψιν, ὅταν συμφωνῇ τῇ ἀληθείᾳ, *In de Part. An.* (ed. Hayduck), p. 15, line 12.

20. Peck translates: "Each of these groups is marked off by *many differentiae,* not by means of dichotomy." Balme (1972, 14) has: "Each of these has been marked off by many differentiae, not dichotomously." Louis *(PA* [1956]) adopts a similar interpretation.

21. 2.642b19. It seems that here πολύπους indeed designates the octopus, as Louis thinks, and not, according to its etymological sense, all the animals with many feet, as most interpreters believe (e.g., Peck, the Latin translation of the Firmin Didot edition, and Balme. If these latter were right, Aristotle's argument would be irrelevant, because there would not really be a splitting of kinds. In any case, it is the usual sense of the word in the biological corpus (see passages cited in Louis's index, 1973, p. 107). Nevertheless, it is certainly strange to see Aristotle imagining that some species of octopus might be terrestrial, for even if an octopus may survive a considerable period of time out of water, it cannot be called amphibious. However, Aristotle is here taking up an Academic example, not one of his own; the octopus may be

thought in some sense to be a "land animal," in that at least some kinds of octopi *walk* with their several feet (on the sea bottom).

22. σύνδεσμος not only means "conjunction" in the grammatical sense of the word, but designates bonds of all kinds. In biology, for example, it can designate the bond that makes the marrow continuously one throughout the articulated segments of the dorsal column (in animals with spiny skeletons, *PA* 2.6.652a16); or the articulation whose flexing permits animals to move from place to place (*IA* 13.712a2). In an entirely different context, one finds this remarkable use in *EN* (8.14.1162a27): "Children seem to be a σύνδεσμος; that doubtless explains why childless couples separate more easily."

23. Eta 6.1045a12. Aristotle repeats the same example of the *Iliad* several times; see *Metaph.* Zeta 4.1030b9; *APo.* 2.10.93b36. It is interesting to note that Aristotle combines the difficulties tied to the problem of the unity of the thing defined with difficulties in the theory of ideas. Thus, the passage of the *Metaphysics* quoted above continues: "What, then, is it that makes man one; why is he one and not many, e.g., animal + biped, especially if there are, as some say, an animal-itself and a biped-itself? . . . Clearly, then, if people proceed thus in their usual manner of definition and speech, they cannot explain and solve the difficulty" (Eta 6.1045a14, trans. Ross). Thus dichotomy, a latecomer in the Platonic philosophy, inherits the aporias of the theory of ideas.

24. A. L. Peck modifies the text of the passage on "the difference of the difference." He writes, ἐὰν δὲ μὴ διαφορᾶς λαμβάνῃ τις (instead of τὴν) διαφοράν, and translates, "but if they do not take the *differentia* of the *differentia*. . . ." It is not obvious why one should modify the text thus.

Since these lines appeared in French (1982), I have adopted a somewhat different way of posing the problem of the unity of the definition in biology. In fact, in a paper presented at Williamstown in July 1983, "Aristotle: A Zoology Without Species" (in Gotthelf [forthcoming (b)]), I have tried, taking up more systematically an idea that was already present in the original version of this work, to show that Aristotle's method of definition fundamentally sought to define the *parts* or *functions,* and not animal species. In contrast, the dichotomic method did try to define species. On this point, too, the Platonic and Aristotelian methodologies are not in competition.

25. D'Arcy Thompson claims that "It would appear that Aristotle's work in natural history was antecedent to his more strictly philosophical work" (preface to his translation of *HA*, p. vii). Thompson's opinion of the dating of the biological works is still arguable to the extent that his remark is based on some positive evidence, the geo-

graphical origin of the animals discussed. However, the phrase "strictly philosophical works" hardly has a meaning when applied to Aristotle.

26. There is almost unanimous agreement that *GA* is a late work. We shall see below that this work represents itself as the conclusion of the general project of Aristotelian biology. But if this work is the last, or nearly so, of the biological writings, does that mean that it was one of Aristotle's last works? Nuyens (1948, 256–63) thought so because he found in it a doctrine of the soul that goes beyond the dualism of soul and body. But to be convinced by his argument, one would have to accept the interpretation that he gives of Aristotle's development on the subject of the soul. The fragility of Nuyens's conclusions has been shown several times, notably by Charles Lefèvre (1972).

I must touch again here on the question of the chronological location of Aristotle's biological work. D'Arcy Thompson thought that the entire biological series was composed in the middle of Aristotle's life, between his two Athenian periods. Thompson supported his argument largely by examining the names of the places of origin of the animals studied, which tends to show that the data included in these books were gathered in northern Greece and Asia Minor. Werner Jaeger (1948), in contrast, maintains that after his metaphysical speculations, Aristotle came at the end of his life to occupy himself with "positive" studies, principally biological. It has been shown by several interpreters that Jaeger's thesis is insupportable: see, e.g., Henry Desmond Pritchard Lee (1948). Nevertheless, Thompson's position also must be criticized. On the one hand, his study is based only on *HA*, yet he extends it to the entire biological corpus; on the other hand, the geographical references gathered by Thompson surely prove that Aristotle carried on some zoological research in northern Greece and Asia Minor. But Thompson does not prove more than that. The most plausible hypothesis is therefore that Aristotle carried on his biological research during his entire life, and thus simultaneously with the rest of his philosophical activity. That is the position of Anthony Preus (1975, 47): "The two parts of his work [philosophical and biological] were not compartmentalized, but rather, the philosophical interest helped to guide the direction of biological investigation, and the biological investigations give insights into the metaphysical problems."

27. Although our discussion here has been of Lloyd's 1961 article, his more recent work appears to continue the same presupposition: see Lloyd (1983, 204–205). This has also been noted by Robert Parker (1984, 184) in his review of Lloyd's book.

28. See *Sophist* 218c1: νῦν γὰρ δὴ σύ τε κἀγὼ τούτου πέρι τοὔνομα μόνον ἔχομεν κοινῇ, "At present you see that you and I possess in common the name" (trans. Cornford); and farther on, the imitation that rests on opinion, "doximimetics" (267b10).

29. *PA* 1.2.642b6: Τοῦτο δ᾽ ἐστὶ τῇ μὲν οὐ ῥᾴδιον, τῇ δὲ ἀδύνατον. This is not a disjunction, as Louis thinks, translating: "Or, ce procédé est tantôt difficile à utiliser, tantôt impraticable" ("Sometimes this procedure is difficult to implement, sometimes impossible"). Rather, the sentence contains a double criticism; both aspects are valuable because each operates on a different level: from a practical point of view, dichotomy is inconvenient, repetitive, and definitely not worth as much as the ordinary way of proceeding to distinctions; but it is also, from a theoretical point of view, an impossible path, in that it never obtains a correct definition, and thus fails to achieve its goal.

30. See especially Guthrie (1957–58); Suzanne Mansion (1961); and Pellegrin (1981).

31. Thus, oppositions that contemporaries have experienced as radical antagonisms may be perceived by interpreters of another day as family quarrels within an implicit intellectual frame of reference. Marsiglio Ficino wrote: "Errant omnino, qui Peripateticam disciplinam Platonicae contrariam arbitrantur" ("Anyone who thinks that Peripatetic teaching is contrary to Platonic is in error"). To be sure, that was a matter of protecting with the authority of Thomistic Aristotelianism the resurgence of a Platonism that disturbed orthodox vigilance.

32. See the references given in the Bonitz *Index*, 524b55–525a11. That definition is not a simple expression is shown by, among other passages, this one from the *Posterior Analytics*: "Since, therefore, to define is to prove either a thing's essential nature or the meaning of its name, we may conclude that definition, if it in no sense proves essential nature (τί ἐστιν, i.e., of the things which one wants to define), is a set of words signifying precisely what a name signifies. But that would be a strange consequence (ἄτοπον); for both what is not substance and what does not exist at all would be definable, since even non-existents can be signified by a name" (*APo.* 2.7.92b26, trans. Mure).

33. See, e.g., *APo.* 1.4.73b3, and generally all of Chapter 4, which presents the various senses of "per se"; *Metaph.* Delta 9.1018a1.

34. συμβεβηκότα καθ᾽ αὑτά: see, e.g., *APo.* 1.22.83b19; *Metaph.* Mu 3.1078a5.

35. E.g., Hamelin (1920, 113): "But it is not only the accident, i.e., the nonessential predicate, that is not constitutive of the concept; it is even the predicate that, based in the essence, nevertheless does not belong to the essence. For Aristotle distinguishes clearly the derived predicates from the essence." And then, in a note, Hamelin cites our passage from the *Metaphysics* (Delta 30.1025a30) in which Aristotle speaks precisely of "per se accidents." Similarly, Edward Seymour Forster translates *APo.* 1.7.75b1 (τὰ καθ᾽ αὑτὰ συμβεβηκότα) by "the essential attributes."

36. A use of the term at 1.639a18, which seems at first glance

different, actually reduces to the sense "per se accident," i.e., necessary but not belonging to the essence: it is necessary to study, says Aristotle, τὰ κοινῇ συμβεβηκότα πᾶσι, "the properties that belong to all animals in common."

37. 645b1 et seq.: 'Αναγκαῖον δὲ πρῶτον τὰ συμβεβηκότα διελεῖν περὶ ἕκαστον γένος, ὅσα καθ' αὐτὰ πᾶσιν ὑπάρχει τοῖς ζῴοις, μετὰ δὲ ταῦτα τὰς αἰτίας αὐτῶν πειρᾶσθαι διελεῖν. A. L. Peck translates the first phrase: "First of all, our business must be to describe the attributes found in each group; I mean those 'essential' attributes which belong to all the animals, and after that to endeavour to describe the causes of them."

38. The understanding of the word συμβεβηκότα is, in any case, as much obscured by translating it "accidents" as "attributes"; we should never lose sight of the first sense of συμβαίνειν, "to go together."

The "parts" are "per se accidents" because they belong to all animals without entering into their *ousia,* in that the *ousia,* as we shall see again below, is nutrition, sensitivity, and movement. If, like Alcibiades, we cut the tail off our dog, it remains a dog; but, as Aristotle often says, a dead animal cannot be said to be the same as before death except homonymously.

39. The last phrase reads, οὐτὰ πάθη καὶ τὰ καθ' αὐτὰ συμβεβηκότα δηλοῖ ἡ ἀπόδειξις. Cf., among others, *Metaph.* Beta 2.997a19: εἴπερ πᾶσα ἀποδεκτικὴ περί τι ὑποκείμενον θεωρεῖ τά καθ' αὐτά συμβεβηκότα ἐκ τῶν κοινῶν δοξῶν, "for every demonstrative science investigates with regard to some subject its essential attributes, starting from the common beliefs" (trans. Ross).

40. It is not part of my purpose to ascertain what "per se accidents" *are.* On this point, one may refer to the analyses of Wolfgang Kullmann (1974), especially p. 182, n. 57, and pp. 261ff. It is certain that συμβεβηκότα καθ' αὐτά have no *accidental* character, in the sense of that word in modern languages. They are, on the contrary, necessary attributes. The pertinent opposition is placed between the characteristics that belong to the *ousia* and the συμβεβηκότα καθ' αὐτά. Thus, for a triangle to have three sides or for some animal to have blood or lungs is a characteristic of its very *ousia* (see *PA* 4.5.678a34; 3.6.669b12).

41. We shall see that the "true" difference, i.e., the only one that ought to be operative for the diairetician, is the "specific difference." In several instances, Aristotle warns against "false" differences; in all cases, they are "false" because they are too immediate. Thus, in *HA* 8.30.608a5: "The eels that are called females are the best for the table: they look as though they were female, but they differ only at first sight (ἀλλὰ τῇ ὄψει διαφόρους)."

42. Pouchet (1884, 365). Pouchet also rests his opinion on an argu-

ment that is the inverse of the one he believes he got from the specialization of the exposition: "We find in it [the long passage on bees], in
the first place, some hesitations, which the least attention of an obser-
vative spirit like his would have sufficed, it seems, to remove; for
example, to know whether such and such a wasp has a sting or not."
We shall have to come back to this sometimes polemical question of
Aristotle's observational talents. For a rapid but adequate review of the
written and oral sources of Aristotelian biology, see Preus (1975, 23–
40).

43. Aristotle speaks often of beekeepers (οἱ μελιττουργοί): *HA*
5.22.554a22; 9.40.623b19, 31; 626a1, 33, b3, 14; 627b6, 13, 16, 19.
But he speaks about the breeders of numerous other animals too:

- the ἐγχελυστρόφοι, eel-feeders: *HA* 8.2.592a2; fragment 294,
  1529a25;
- the φαρμακοπῶλοι who raise snakes and tarantulas (*HA* 8.4.594a23)
  or only spiders (*HA* 9.39.622b34);
- the pork-breeders, called either ὑοβοσκοί from the name of the
  animal (ὕς) (*HA* 8.21.603b5), or οἱ πιαίνοντες, "the fatteners," *HA*
  8.6.595a25);
- the ποιμένες who raise sheep (*HA* 6.18.573a2, 8, 10; 8.10.596a30;
  9.3.610b28);
- the mahout, ὁ ἐλεφαντιστής (*HA* 2.1.497b28; 9.1.610a27), and the
  camel-driver, ὁ καμηλίτης (*HA* 9.47.630b35).

This list of animal-raisers is far from exhaustive. For domestic animals,
see Louis (1970).

44. What Aristotle says about fishermen (οἱ ἁλιεῖς) is more likely
to suggest an indirect investigation than an actual participation in fish-
ing expeditions: they pull up their nets at dawn or dusk (*HA*
8.19.602b9), they approach schools of fish silently (*HA* 4.8.533b29),
they notice where the silurus is watching over its eggs (*HA*
9.37.621a28); all that could be learned in port. On this point, see the
relevant remarks by Friedrich Solmsen (1978), especially p. 474. Aristotle speaks of "experienced" (ἐμπειρικοί) fishermen who say that
they have seen animals resembling sticks of wood in the shape of the
male sexual organ (*HA* 4.7.532b20), doubtless to signify that this information should be taken into consideration. Beside the error noted
about the reproduction of fish, Aristotle blames fishermen for believing
that octopi couple by means of a tentacle (*GA* 1.15.720b32), which *is* in
fact their means of sexual congress. For the description of the coupling
of octopi, see *HA* 5.6.541b1 et seq.

45. Perhaps we have here, from Aristotle's pen, an indirect definition of ἱστορία, observation from a theoretical point of view although

not yet attaining the general. Louis writes (1955b, 44): "The same difference [as that which exists between history and poetry] separates ἱστορία, knowledge, from ἐπιστήμη, science: the latter can only deal with the general, while the former is satisfied with gathering particular facts with all possible exactitude."

46. "Herodotus the fabulist," Ἡρόδοτος ὁ μυθολόγος, says Aristotle (*GA* 3.5.756b6). We must agree that Aristotle was not able to recognize the scientific value, particularly in its methodological plan, of Herodotus's inquiry. In a rich and brilliant article on Aristotle's use of the story of a torture that the Tyrrhenian pirates inflicted on their prisoners, tying them alive to cadavers, Jacques Brunschwig (1963, 175) reminds us that Aristotle was not always "the positive ethnographer, but the amateur of myths, the φιλόμυθος, which he had been at the beginning of his life and which he became again in his old age." Brunschwig cites in a note the passage of the *Metaphysics* (Alpha 2.982b18) in which Aristotle says that to be a lover of stories is already, in a way, to be a lover of wisdom.

47. It seems that there are in Aristotle two distinct attitudes about observations done by others than himself, related to two sorts of facts.

There are, in the first place, more or less fabulous or extraordinary facts that generally provoke his skepticism; Louis (1967) presents these very well. Thus, in the case of the martichoros (*HA* 2.1.501a25), Aristotle leaves Ctesias with the responsibility for the description; and we know how little confidence he has in "Ctesias, no very good authority" (*HA* 8.28.606a8), of whom he notes elsewhere that he has been in error also about the semen of elephants, which he says becomes, upon drying, something like amber (*GA* 2.2.736a2; *HA* 3.22.523a26). It is not without interest to notice that Aristotle's successors in the study of animals, although they did not hesitate to pillage him, most often did not do it with his own critical spirit. Thus, Aelian gives as verified the fable which says that cranes swallow a stone for ballast and then vomit it again, and that this stone then becomes a touchstone for testing gold (*Nat. An.* 3.13), while this fable had been explicitly denounced as such by Aristotle (*HA* 8.12.597b1).

Also, there are the facts that seem probable to Aristotle but that he has not checked. Doubtless that is the case here, with the sensitivity of sponges.

Sometimes Aristotle notices that opinions diverge in his sources, and he gives two theses without choosing between them. We have already seen this way of dealing with things in the case of the "wild man." In *HA* 2.2.501b5, he notes that some think that dogs lose their teeth, and others think they keep them.

Lacking precise observations, Aristotle sometimes suspends judgment. For example, the question of whether the nautilus can live without its shell has not yet been decided (*HA* 9.37.622b18); it is not known whether the queenbee (the king, according to Aristotle) is to be found among the bees that swarm around the hive before hiving off (*HA* 9.40.625b11); satisfactory observations have not yet been made on the old age of wasps and of their queen, and on the destiny of the queen after reproduction (9.41.628a25); "it seems that wasps lose their sting in the winter, but we have not met any eyewitness" (628a7). I have given these last examples to show the importance of practical men: Aristotle misses, in the case of wasps, the group of observations gathered by beekeepers on bees.

Nevertheless, Aristotle accepts some curious facts that he has certainly not verified. His detractors cite some of them, almost always the same ones, which proves that they copy from each other. Even though it is rarely invoked, I must confess that I have a hard time believing that lobsters die of fear when they sense the presence of an octopus in the net that has captured them (*HA* 8.2.590b15). But, after all, I have not verified the matter either.

The importance of recourse to "they say" will appear more clearly in the light of a count that I have made: in the last two books of the *History of Animals,* which should make more appeals to tradition because they are devoted to the habits of animals — that is, to facts that are difficult to verify by any one person, or even by one group of researchers — there are thirty-eight references to facts reported by others. Aristotle calls attention to these references most often with the word φασι but also by λέγεται or λέγουσι.

48. I have, in this study, made the case only for the biological texts. But elsewhere (Pellegrin [1981]) I have also discussed logical and metaphysical texts relating to division, particularly *APr.* 1.31 and *APo.* 2.5.

49. LeBlond (1939b, 359). I must, to forestall possible criticisms, say about definition what I have already said about "per se accidents": there is no question of treating it for its own sake here, even in the narrow frame of Aristotle's biology. There are indeed several kinds of definition in Aristotle, at least three according to the *Posterior Analytics* (2.10). As for the *purpose* of definition in biology, I have treated it more exactly elsewhere ("Aristotle: A Zoology Without Species," in Gotthelf (forthcoming [b]).

2.   GENUS, SPECIES, AND SPECIFIC DIFFERENCE: FROM LOGIC TO ZOOLOGY

1. LeBlond (1939a, 68, 71, 72). Even Georges Canguilhem, to be sure with the much greater prudence of one who has the confused feelings of going forward on unsure ground, leans toward this interpretation when he writes (1968, 336): "The theory of the concept and the

theory of life — are they not of the same age, and by the same author? And does not this same author tie them to each other at the same source? Is not Aristotle simultaneously the logician of the concept and the systematician of living beings?"

2. Rey (1946, 153). Léon Robin (1944, 66–67) is also of this opinion, even if he states it less clearly, resting it on chronological considerations: "Outside the zoological studies of Assos, which belong to the first period of his career and to the Academic practice of the method of 'division,' the properly scientific works of Aristotle, I mean those that deal with facts, were not undertaken until much later, when the idea of what science *ought* to be had already been solidly structured in his thought." Pierre Louis (1975, 19) has a more balanced reading: "He [Aristotle] will find in natural history a remarkable field of application for logic, at the same time as he will apply in biology the new methods of the logician. It would be useless to ask which of these two sciences preceded the other in the development of Aristotelian thought: simply put, there has been between these two a lasting and fruitful interaction."

3. Julius Stenzel (1929) seems to think so, though David Balme (1962a) supports the opposite position. But, as Marx and Engels say in the *German Ideology,* in the absence of facts, speculation "can loosen the bridle of its speculative instinct and can engender and give birth to hypotheses by the thousands." Louis Bourgey (1953) underlines, first, that the word *eidos* does not have in the Hippocratic corpus the technical sense that the fourth-century philosophers would give it (p. 34, note); then he cites a passage from the *On Food* (ed. Émile Littré, vol. 9, p. 98, no. 1) in which one finds a use comparable to Aristotle's of *genos* and *eidos* as "genus" and "species" (p. 39, note 4). The Hippocratic text is the following: τροφὴ καὶ τροφῆς εἶδος, μία καὶ πολλαὶ μία μὲν ἦ γένος ἕν, εἶδος δὲ ὑγρότητι καὶ ξηρότητι. In fact, what I will show for Aristotle will also work for this Hippocratic text, possibly contemporary with his work; and in this sense Bourgey is right, although he is mistaken in thinking the pair *genos/eidos* equivalent to the pair "genus/species," thus sharing the common prejudice of Aristotle's interpreters.

4. Balme (1962a, 82): "The fact that he [Plato] did not find it necessary to create a verbal distinction between genus and species is perhaps a straw on the side of those who hold that he did not use diairesis for systematic classification," a group to which, obviously, I belong. Janine Bertier writes (1977, 62): "The disparity of appellations of groups of animals in the *Timaeus — genus, race, people —* is surprising, but it is not unique. We notice the equivalent way in which the Hippocratic author of *Airs, Waters, Places* denominates the peoples whose characters he describes. In the *Timaeus,* this disparity is not very serious,

because the ordering of animals is based on a cosmic localization, and no formal logical system is applied to the homogenization of these terms in positing them as genera and species. With Aristotle, the equivalence of the terms *race, people, genus* will disappear in favor of a system of classifying the living forms treated, before all and as much as possible, in terms of genus and species." As much as I agree with the first part of this interpretation — the part about Plato — I must reject the last part (that, among other things, is the goal of this chapter), a thesis that Bertier has not *established,* but has gathered from the set of (mis)presuppositions of the Aristotelian tradition.

5. *Mete.* 1.1.338a20 (trans. Lee, Loeb edition): "We have already dealt with the first causes of nature and with all natural movement; we have also dealt with the ordered movements of the stars in the heavens, and with the number, kinds and mutual transformations of the four elements, and growth and decay in general." Even if Aristotle's allusions are not always to finished treatises, or to treatises in the form in which we know them, this introduction shows the relative lateness of the *Meteorologica.*

6. Aristotle denotes what we call "specific difference" by several words or expressions: by the word διαφορά in a technical sense that he gave it; by the verb διαφέρειν, which is most often specified by the dative εἴδει; and by expressions such as τὸ ἕτερον τῷ εἴδει, which indicate an alternativity according to the εἶδος. (See, for example, *Metaph.* Iota 7.1057b35; *HA* 2.1.497b10.) Other expressions include: διαφέρον εἴδει (e.g., *Metaph.* Iota 3.1054b27; *HA* 2.15.505b35) and διαφέρον κατ' εἶδος (e.g., *Metaph.* Delta 10.1018a31). Elsewhere Aristotle gives for equivalents the specification of the *genos* by the specific difference, and the qualification of the substance, for the specific difference is to the *genos* what quality is to substance: τὸ ποιὸν λέγεται ἕνα τρόπον ἡ διαφορά τῆς οὐσίας, οἷον ποιόν τι ἄνθρωπος ζῷον ὅτι δίπουν, "in one way, quality means the specific difference of the substance; thus man is an animal that has a certain quality, i.e., biped" (*Metaph.* Delta 14.1020a33). One finds this doctrine again in the *Topics,* where it is referred to several times: οὐδεμία γὰρ διαφορὰ σημαίνει τί ἐστιν ἀλλὰ μᾶλλον ποιόν τι, καθάπερ τὸ πεζὸν καὶ τὸ δίπουν, "no specific difference signifies the essence but rather a certain quality, as terrestrial or biped" (*Top.* 4.2.122b16; cf. 6.6.144a18). Indeed, posing the question of quality for Aristotle sometimes amounts to searching for the specific difference. Here are two examples. In the *Nicomachean Ethics,* after having established that excellence (ἀρετή) belongs to the class of habits (ἕξεις), Aristotle writes: "That which is thus excellence according to its *genos,* we have just said. But we must not only say that it is a habit, but also what sort (*poia*)" (2.5.1106a13).

Here the quality thus specifies the *genos* "habit." In the *Parts of Animals,* after having criticized the "old philosophers," Aristotle enumerates the questions that one ought to ask oneself: "One ought to say that the animal is *such,* and to speak about that — what it is, its qualities, and those of each of its parts" (1.1.641a15). It does seem that the two questions marked by "both what and of what sort" are two steps of a specification, the first inviting the determination of the *genos* of the animal (and perhaps of each of the parts of the animal, but it is still too early to begin that debate), and the second specifying by specific difference. The fact that the *genos* should be specified by the specific difference as the substance is "modified" by the quality already takes us away from any taxonomic perspective in which the pair *genos/eidos* would function like "genus/species." There is a related distinction in a famous passage of the *Generation of Animals* (4.3.767b32): γεννᾷ δὲ καὶ τὸ καθ' ἕκαστον τοῦτο γὰρ ἡ οὐσία καὶ τὸ γινόμενον γίνεται μὲν καὶ ποιόν τι, ἅμα δὲ τόδε τι καὶ τούτη ἡ οὐσία, "it is simultaneously the individual and the *genos* that generate, but primarily the individual, since it is the *ousia*. And the product of generation is simultaneously the product of a certain 'quality' and an individual entity, and this is the *ousia*." Here ποιόν τι, "certain quality," designates a class that, in this particular case, we would denominate a species, while *ousia* designates a concrete individual.

7. Iota 3.1054b27: πᾶν γὰρ τὸ διαφέρον διαφέρει ἢ γένει ἢ εἴδει, γένει μὲν ὧν μή ἐστι κοινὴ ἡ ὕλη μηδὲ γένεσις εἰς ἄλληλα, οἷον ὅσων ἄλλο σχῆμα τῆς κατηγορίας, εἴδει δὲ ὧν τὸ αὐτὸ γένος.

8. Iota 4.1055a6: τὰ μὲν γὰρ γένει διαφέροντα οὐκ ἔχει ὁδὸν εἰς ἄλληλα, ἀλλ' ἀπέχει πλέον καὶ ἀσύμβλητα.

9. It seems that the term is not found in Plato. The hypothesis that seems most probable to me is that Tricot attributes it to Plato as a consequence of a bad interpretation of Bonitz, whom Tricot read assiduously. On *Metaphysics* Mu 6, where a distinction is made between ideal numbers and mathematical numbers, Bonitz writes (1849, 540): "For despite the fact that Plato, according to Aristotle's testimony, seems to have said that *the ideas are numbers,* nevertheless he distinguished these numbers, which express the nature of the ideas, from mathematical ones by this reason that in fact entirely removes from them the very nature of numbers. For he told *(dixit)* us that [these two kinds of] numbers are ἀσύμβλητους, can be neither associated nor computed with each other." Doubtless, Tricot believes that *dixit* in the last sentence has "Plato" understood as subject, but it is obviously an assertion of Aristotle talking about Platonism. As we have seen on the subject of the term διχοτομία (see note 16 in Chapter 1), Aristotle here accounts for Platonic thought in non-Platonic terms. The closest textually Pla-

tonic notion seems to be that of ἀσύμμετρος. See the remarks of Charles Mugler (1956).

10. Iota 3.1054b30: λέγεται δὲ γένος ὃ ἄμφω τὸ αὐτὸ λέγονται κατὰ τὴν οὐσίαν τὰ διάφορα.

11. Iota 8.1057b37: τὸ γὰρ τοιοῦτον γένος καλῶ ὃ ἄμφω ἓν ταὐτὸ λέγεται, μὴ κατὰ συμβεβηκὸς ἔχον διαφοράν.

12. Iota 8.1058a7: λέγω γὰρ γένους διαφορὰν ἑτερότητα ἢ ἕτερον ποιεῖ τοῦτο αὐτό.

13. This problem is often treated by Aristotle, notably in the *Posterior Analytics* (1.7.75a38): "There is no demonstration when one passes into another genus" (ἐξ ἄλλου γένους μεταβάντα). Cf. *APr.* 1.30.46a17; *Top.* 8.12.162b7; *Metaph.* Alpha 9.992a17, Delta 28.1024b15.

14. See, for example, among recent works, Louis (1978).

15. We can locate these hitches by means of three terms:

- διαστρέφειν in the first sense, which means "turn oneself": the talons of eagles turn themselves (*HA* 6.6.563a24); for the eyes: to turn the eyes, be cross-eyed (*EE* 8.1.1246a29; *Probl.* 31.7.957b36, 27.960a9). This term subsequently comes to designate a distortion in that which should have been according to nature; thus, concerning the triton (water-newt, κορδύλος), Aristotle writes: "It seems that nature has undergone a kind of deviation [turning]," ἡ φύσις ὡσπερανεὶ διεστράφθαι (*HA* 8.2.589b29).

- ἐπαμφοτερίζειν means "to hold the middle between two groups." Aristotle sometimes uses it in a properly classificatory way, but not taxonomically, in making it designate animals that can be *indifferently* ranked into two different groups. Thus, in *HA* 8.2.589a21, on animals that, although breathing air and giving birth on solid ground, spend their life in an aquatic milieu: ἅπερ ἔοικεν ἐπαμφοτερίζειν μόνα τῶν ζῴων καὶ γὰρ ὡς πεζὰ καὶ ὡς ἔνυδρα τις ἂν θείη, "It seems that alone among animals they belong to two classes, because one can as well rank them with the terrestrial animals and among aquatic animals." Plato uses the term in the sense of "be unsure, ambiguous" (e.g., *Rep.* 5.479b11) or "to be shared between two parties" (*Phdr.* 257b6). In Aristotle's biology, the most frequent use is to designate animals that belong to an intermediate group: for example, monkeys (*PA* 4.10.689b32) and seals and bats (*PA* 4.13.697b1). The fact that this term functions on the edges of *gene* is brought out well at *PA* 4.5.681a36: "As for what are called Knides or Akalephae, they are not Testaceans, but they fall outside the defined *gene*, and they incline simultaneously to the nature of the plant on the one side and to that of the animal on the other," ἀλλ' ἔξω πίπτει τῶν διῃρημένων γενῶν, ἐπαμφοτερίζει δὲ τοῦτο καὶ φυτῷ καὶ ζῴῳ φύσιν.

• ἐπαλλάττειν (or the noun ἐπάλλαξις) indicates "crossing," in the first sense of the word: it is used of bodies whose parts chain together, that is to say, whose links overlap (*Mete.* 4.9.387a12), or to designate the action of crossing one's fingers (*PN* 460b20; *Metaph.* Gamma 7.1011a33). In biology, the expression ἐπάλλαξις τοῖς γένεσιν (*GA* 2.1.732b15) indicates the intersection of different groups (bipeds, quadrupeds, vivipara, ovipara). In *GA* 4.6.774b17, it is said that the pig treads on two families, or, more exactly, reunites two characters usually incompatible in being multiparous and hooved. To give an idea of the diversity of use of this verb, here are a few references, taken from the *Generation of Animals:* at 1.14.720b10, it serves to describe the coupling of crustaceans; at 4.4.769b34, it is said that, according to Democritus, monsters come from the fact that two seeds have penetrated the uterus "so that the parts [of the embryo] grow together and cross each other," ὥστε συμφύεσθαι καὶ ἐπαλλάττειν τὰ μόρια; at 770b6, still about monsters, Aristotle says that they are more frequent in animals that give birth to imperfect young — "that is why this accident ἐπαλλάττει to animals of this kind." (Pierre Louis has: "se produit par intermittences"; A. L. Peck has: "extends its range to.")

16. *Metaph.* Iota 3.1054b23: διαφορὰ δὲ καὶ ἑτερότης ἄλλο. τὸ μὲν γὰρ ἕτερον καὶ οὗ ἕτερον οὐκ ἀνάγκη εἶναι τινὶ ἕτερον· πᾶν γὰρ ἢ ἕτερον ἢ ταὐτὸ ὅ τι ἂν ᾖ οὐ· τὸ δὲ διάφορον τινὸς τινὶ διάφοραν, ὥστε ἀνάγκη ταὐτό τι εἶναι ᾧ διαφέρουσι.

17. We find this remarkable statement in the *Topics* (6.6.143b8): πᾶσα γὰρ εἰδοποιὸς διαφορὰ μετὰ τοῦ γένους εἶδος ποιεῖ; the term εἰδοποιός, which is rare but expressive (Bonitz cites only one other use, in *EN* 10.3.1174b5; one finds the term again in the Neoplatonists), insists here, in a somewhat pleonastic fashion (εἰδοποιὸς . . . εἶδος ποιεῖ), that it is the specific difference that constitutes the *eide* within the *genos;* one may translate: "for every *eidos*-producing difference produces *eidos* in adding it to the *genos.*"

18. See Daudin (1926, 66ff.).

19. Hamelin (1920, 141). It would be more exact to say that hierarchy and filiation are superimposed, which could be expressed by the word "descent," meaning "coming from," suggesting a position at a lower level. To the best of my knowledge, the best general study of contrariety in Aristotle remains John Anton's *Aristotle's Theory of Contrariety* (1957), which restores the Aristotelian doctrine to its historical and theoretical context, with, however, perhaps too little attention to

biology. The problems that interest us here are treated especially in Chapter 6.

20. We find in the *Topics* (6.6.143a36) a somewhat different expression: πᾶν γὰρ γένος ταῖς ἀντιδιῃρημέναις διαφοραῖς διαιρεῖται, "every *genos* is divided by differences obtained by divisions into contraries"; and Aristotle gives as example the division of the *genos* by these differences: pedestrian, winged, aquatic, biped. The term ἀντιδιῃρημέναις indicates that the differences in question have been found at the end of a division.

21. I can give further support to this interpretation by considering the passage in the *Metaphysics* (Delta 10.1018a25–b8) that provides two successive lists of the different senses of ἐναντία and of ἕτερα τῷ εἴδει.

I must call attention to at least one case in which Aristotle breaks his own rules of distinction. In the *Ethics,* he distinguishes three *eide* of friendship, according to whether it is based on the good, pleasant, or useful. We are dealing here not with three notions, but with three pairs of contraries gathered from what we could call the "philosophical tradition" of his time (see *Topics* 1.13.105a27, 3.3.118b27). In the *Nicomachean Ethics,* he associates these three notions with their contraries by relating them to the opposition of choice and refusal. In all rigor, these three notions cannot be called *eide* of the same *genos.* Nevertheless, Aristotle declares that they "differ from each other according to *eidos*," διαφέρει δὲ ταῦτα ἀλλήλων εἴδει (*EN* 8.3.1156a6). I think we are dealing here with a "sliding" from the notion of contraries to the notion of specific difference, entirely admissible in a work so scientifically unrigorous as the *Ethics*. But this slide is not without consequences in this passage, as in the parallel passage in the *Eudemian Ethics.* Let us look at the two texts.

"Is there one *eidos* of friendship, or are there several? For those who believe that there is only one, because it admits of the more and the less, rely on an insufficient proof; for the things other according to *eidos* are also susceptible of the more and the less" (*EN* 8.2.1155b12). The objection to which this passage responds is all the more interesting because it seems to emanate from an "orthodox" Aristotelian who based it on, for example, the doctrine (noted below) according to which quantitative variations occur within the *eidos* (see also *Politics* 1.13.1259b37, just cited, which opposes specific difference and quantitative variation). Tricot, in a note to his translation, claims that Aristotle means that the more and the less can be considered as "species." Gauthier and Jolif write in their commentary (1959, 669) "that one ought not to understand the word "species" (εἶδος) in this passage in the technical sense of species of the same genus." In my opinion,

Aristotle is simply warning here against a confusion: to say that *eide* admit the more and the less does not mean that contraries admit intermediates (see note 22, below). Thus, health is not intermediate between illness and excellent health. I shall take up in more detail below the problem of the relations of *eidos, genos,* and the more and the less.

"It is thus necessary that there be three *eide* of friendship and that neither all the friendships be said to be of the same *[eidos],* nor that these *eide* be called *eide* of a single *genos,* nor that they be entirely homonymous [equivocal]" (*EE* 7.2.1236a16). The fact that there would be three *eide* of friendship is deduced from the fact that the three ends on which friendship can be based—the good, the pleasant, and the useful—have been called *eide.* But Aristotle warns the reader that here the logical schema of the division of *genos* into *eide* does not apply. He seems himself to corroborate the interpretation of such texts as theoretically approximate (see *EN* 1.1.1094b13).

A closer examination of the texts will show that in the *Ethics* Aristotle does not mean to bring into action specific difference in its most rigorous sense for delimiting the *eide* of friendship. Thus, in the *Eudemian Ethics,* where he often repeats that there are three *eide* of friendship, he summarizes what has been said in the first chapters of Book 7 as follows: "We have thus said that there are several sorts (τρόποι) of friendship, and how many there are, i.e., three" (7.5.1239b3). It is symptomatic that Aristotle gives the term τρόπος as equivalent to the term *eidos,* when it is vague and cannot be a subdivision of a true *genos.*

22. For Aristotle makes clear that certain pairs of contraries admit of intermediates and others do not: "Those contraries which are such that the subjects in which they are naturally present, or of which they are affirmed, must necessarily contain one or the other of them, have no intermediate between them; but those in which no such necessity obtains, always have an intermediate" (*Cat.* 10.11b38, trans. Edghill, Oxford edition).

23. See the passage of the *Categories* cited by Ross (13.14b33): "That which is obtained by division within the same *genos* by the contraries (τὰ ἐκ τοῦ αὐτοῦ γένους ἀντιδιῃρημένα ἀλλήλοις) is said to be 'simultaneous' in nature. I mean by 'obtained by division by the contraries' the things that come out of the same division (τὰ κατὰ τὴν αὐτὴν διαίρεσιν), as winged for terrestrial and aquatic (οἷον τὸ πτηνὸν τῷ πεζῷ καὶ τῷ ἐνύδρῳ)."

According to my interpretation, we must understand this passage not as showing a triple division (winged/pedestrian/aquatic), but as showing two divisions from "winged" taken as a reference point: that is what the difference of cases in the three terms (a nominative versus two datives) indicates. Similarly, the example that follows the passage in the *Topics* cited above, in note 20, gives conjointly the three terms "pedestrian, winged, aquatic," and adds "biped."

In *PA* 4.13.697b1 et seq., Aristotle makes two pairs of contraries function side by side — terrestrial/aquatic and terrestrial/winged — which again confirms this interpretation.

24. This multivocity of contrariety is precious for the interpretation of Aristotle's moral theory in general, and of his doctrine of "virtue" in particular. For Aristotle, virtue, or rather excellence, is a μεσότης, a term whose least bad translation is perhaps "mean," since Aristotle introduces it from the perspective of mathematical considerations (*EN* 2.5.1106a26 et seq.). Commonly, Aristotle's position has been related to the value of the "right measure" in Greek thought, both philosophical and popular. Several interpreters have remarked that Aristotle found the model for his ethical concept of μεσότης in medicine, or more exactly in the Hippocratic theory of health as an equilibrium and mean (see Wehrli [1951] and, more generally, Jaeger [1957]). It must be noted that when Aristotle analyzes this mean in terms of pairs of contraries (*EN* 2.5.1107a8, 24), he holds conjointly two theses that look incompatible only in rough translations: (1) Virtue is a mean; (2) Virtue is an extreme (ἀκρότης). Right measure, in fact, is not *in the middle* of a division (the one that opposes defect to excess), but at the *extreme* of two divisions, in which it is successively opposed to excess and to defect. This morality of the mean that is extreme, even heroic, is not, as so many people have thought, a morality of conformist moderation.

25. Iota 4.1055a19–21: τούτων δὲ ὄντων φανερὸν ὅτι οὐκ ἐνδέχεται ἑνὶ πλείω ἐναντία εἶναι (οὔτε γὰρ τοῦ ἐσχάτου ἐσχατώτερον εἴη ἄν τι, οὔτε τοῦ ἑνὸς διαστήματος πλείω δυοῖν ἔσχατα). In Greek, διάστημα seems to include not only the idea of separation, which it has etymologically (it is related to διίστημι), but also a spatial nuance, which makes "interval" a good translation; even when it is applied to time or sound, this word invites us to spatialize differences. Employed by mathematicians, the Greek term is often applied to music, starting from the pseudo-Aristotelian *Problemata* (19.23.919b11; 47.922b6); on this subject, cf. Theon of Smyrna, *Music* 3–5 (ed. Dupuis, pp. 80–82).

This spatialization shines clearly through Alexander's commentary: ἄλλως τε πᾶσα ἐναντιότης ἕν ἐστι διάστημα, παντὸς δὲ ἑνὸς διαστήματος δύο τὰ ἔσχατα, ὥστε καὶ πάσης ἐναντιότητος δύο τὰ ἔσχατα (ed. Hayduck, p. 620, line 3). In the multivocity of contrariety that I have just presented, we find, in another form, one of my earlier analyses. In the interpretation that I proposed of *PA* 1.3.643b9 (see pp. 25–31, above), I showed that the true definition is "polygonal," in that the thing defined is reached from several sides at the same time, according to what I there called different "axes of division."

26. See the references given by Bonitz *Index* 378a36–38. The interpretation is notably that of Robin (1944, 99).

27. *Isagoge,* ed. Busse, p. 4, line 15; ed. Brandis, 2a6. γενικώτατον and εἰδοκώτατον are not used by Aristotle.

28. Aubenque (1978, 4): "It would be otiose to show that there is no occurrence of the term *analogia* (or of related terms) in the entire Aristotelian corpus which does not support that definition, and that is the case whether it is rhetorical, ethical, biological, or metaphysical usage of the notion."

29. "'Alteration' occurs when the substratum, which is perceptible, persists, but there is change in its properties, which are either directly or intermediately contrary to one another; for example, the body is healthy and then again sick, though it persists in being the same body, and the bronze is spherical and then again angular, remaining the same bronze" (*GC* 1.4.319b10, trans. Forster, Loeb edition).

30. Aristotle has often been praised for having studied man as an animal. Louis, for example, writes (1975, 101): "Aristotle studies him as an animal in the midst of the others." Louis subsequently notes the preeminent place that man occupies in Aristotle's studies. In fact, this "compliment" for Aristotle has an anachronistic side, in that it can be understood only in the light of later disputes: Darwin would doubtless deserve the compliment, for, against the idea of a theological origin of a separate "human realm," he asserted the animal ancestry of man. I shall have occasion to show that Aristotle's zoology remains fundamentally anthropocentric.

For the "lineage" sense of *genos,* I rely here on one example: when, in the *Politics,* Aristotle writes, ἐν δὲ ταῖς κατὰ γένος βασιλείαις (5.10.1313a10), we must obviously understand "in hereditary royalties." Cf. *Pol.* 3.14.1285a16; 15.1285b39. Can *genos* designate "sex" in Aristotle? That was the case in Plato — for example, in the *Cratylus* (392c8), where this term is applied to the masculine or feminine sex of the human race. The examples given by Bonitz reveal that he has read too rapidly: in *HA* 9.1.608a35, ἐν δὲ ταῖς ἄλλοις γένεσι τὰ θήλεα μαλακώτερα means "in the other species [i.e., bears and panthers, which he was just talking about], the females are softer"; in *HA* 9.40.623b8, the expression γένη . . . μελιττῶν designates the varieties of hymenoptera, of which Aristotle gives a list (in any case, he speaks of "nine varieties," γένη ἐννέα, which absolutely excludes translation by "sex"). The only text cited by Bonitz that would tend in his direction is found in the *De Plantis* 1.1.815a20, which is certainly inauthentic. However, there are two passages in the *Generation of Animals* in which — in the first, probably; and in the second, certainly —

"sex" is designated by *"genos."* At 1.1.715b5, we read: "but those that are born not of animals but of the putrification of matter generate an offspring of another *genos,* in that it is neither female nor male." At 3.7.757a22, Aristotle raises questions about the *"genos* male," τὸ δ' ἄρρεν γένος. On the other hand, Bonitz is quite right in noticing that in the *Rhetoric* (3.5.1407b7) τὰ γένη τῶν ὀνομάτων designates exactly what we call "the [masculine, feminine, and neuter] genders of nouns."

31. Pellegrin (forthcoming). The etymology of these terms, if I may return to that, permits opposition of *genos* and *eidos* in several ways. Etymologically, *eidos* is the visible aspect of a thing, that is, its truth. It was only later that Western thought would oppose the visible aspect to the hidden aspect as the undependable or incomplete appearance of reality. But rather than opposing *genos* and *eidos* as "relationship" and "look," as Balme does, I prefer to see in *eidos* the *immediate* presence of one or more properties of an object, and in *genos* the logical space within which the *process* of its realization occurs.

32. It should be noted that Bonitz indicates uses of the term in this sense only in the ethical and political texts. Among these texts, not counting a reference to the *Rhetoric to Alexander,* at least one does not belong in this section: in *Pol.* 5.3.1303b14, the subject is "differences that provoke oppositions" that cause seditions in cities; but as examples of these differences, Aristotle gives the pairs virtue/vice and wealth/poverty, which shows that there is here a use of *diafora* in his usual sense, and that by giving it the sense "dissensio" Bonitz takes the cause for the effect. The other passages are *Pol.* 5.4.1303b38, which concerns a matrimonial quarrel; 7.16.1334b38: differences between spouses; *Magna Moralia* 2.11.1210a25 (and a35, b1): disagreements between friends; *EN* 1.1.1094b15: disagreements about the good and the just in politics. These passages, perhaps with the exception of the first, which is more "theoretical," are taken from texts that seem to be intended for a larger public; that is doubtless the case with the last two books of the *Politics,* in which Aristotle speaks of the best city, and the ethics, especially the *Nicomachean Ethics.*

33. *Histoire des Animaux,* vol. 1, p. 169, n. 7.

34. Since the appearance of the French version of this work, Balme has recognized the accuracy of my (friendly) criticisms on this point, with honesty and kindness.

35. *Histoire des Animaux,* vol. 1, p. 162, n. 5. Eleven years later, Louis was much more circumspect: "He [Aristotle] tried to express the idea of an order or family by the expression *great genus,* which is found in several passages in the biological treatises. Sometimes it is applied to mere genera, but it can also designate larger groupings, as the entire group of testacea or that of insects" (Louis [1975, 155]).

36. When he asks himself whether the doctrine of difference with

respect to the more and the less *within a genos* is peculiar to the biological corpus, Lennox cites at least one text outside that corpus that seems to allude to it; I think, however, that he has not read that passage correctly. It is a passage from the *Physics* (5.2.226b3), which he translates thus: "within the same genus, but with respect to the more and the less" (the subject is qualitative changes); while the text has ἡ δ’ ἐν τῷ αὐτῷ *εἴδει* (and not γένη) μεταβολὴ ἐπὶ τὸ μᾶλλον καὶ ἧτον ἀλλοίωσις ἐστιν. On the other hand, there is a passage in the *Nicomachean Ethics* (8.2.1155b12) that Lennox could have cited: "Those who think that friendship has only one *eidos* because it admits the more and the less rely on an indication that is inappropriate; for the same things differ according to the more and the less and according to the *eidos*. We have dealt with this problem elsewhere." This passage shows, at least, that the doctrine of the difference of degree within a *genos* had been discussed outside the domain of biology, and that texts, now lost to us, were devoted to it. Nonetheless, in the Aristotelian corpus as we have it, this doctrine seems to have been completely developed only in the biological writings.

37. For this doctrine, see Rorty (1973). Marjorie Grene (1974) argues against Rorty's conclusions in an incisive article. According to her, though one may say that "the concept of *genos* has some matter-like function sometimes in the analysis of nature" (p. 118), this concerns an attenuated sense, "a very metaphorical meaning" (p. 120) of the term "matter."

38. See also *EN* 1.2.1095b2. Note that "prior by nature" can be combined with a chronological posteriority: thus, after having shown that the city is the historical end of an evolution beginning with the family, passing through the village and the monarchical state, Aristotle writes that "the city is prior by nature to the family as well as prior to each one of us" (*Pol.* 1.2.1253a19).

39. LeBlond (1945, 42). This anthropocentrism is very well summarized in the remarkable formula of *IA* 4.706a19: "Man is the animal that most closely conforms to nature." We shall come back to this.

Aristotle's notion that all living things are fundamentally identical has interesting ramifications for him. For example, there is the idea that infinitesimal changes in *quantity* can cause the diversity of animal species: "We may infer, then, that if in the primary conformation of the embryo an infinitesimally minute but absolutely essential organ sustains a change of magnitude one way or the other, the animal will in one case turn to male and in the other to female; and also that, if the said organ be obliterated altogether, the animal will be neither of one sex nor of the other. And so, by the occurrence of modification in minute organs, it comes to pass that one animal is terrestrial and another aquatic, in both senses of these terms. And again, some animals are amphibious and others are not, owing to the circumstance that in their conformation while in the embryonic condition there got intermixed into them some portion of the matter of which their subsequent food is constituted" (*HA* 8.2.590a2). Compare *Pol.* 5.3.1302b34: "For as a body is made up of many members, and every member ought to grow in proportion, that symmetry may be preserved; but loses its nature if the foot be four cubits long and the rest of the body two spans; and should the abnormal increase be one of quality as well as of quantity, may even take the form of another animal" (trans. Jowett). This is not only a "dialectical" notion, by which quantitative change is at the base of qualitative change, but even a possible subversion of Aristotelianism itself. If, in fact, the *eidos* of an animal can depend on the change of size of a part, there will be two formally anti-Aristotelian consequences: first, in some cases, it is the matter, in the last analysis, that determines the *eidos,* and the last phrase of our quotation from *HA* 8.2 gives one such case; second, something quantitative, i.e., numerable (if not measurable), explains by its variations the diversity of reality: there is here a kind of return to a form of Pythagoreanism, a doctrine that Aristotle nevertheless fought.

40. Even though comparative anatomy must be anti-hierarchical if it is to be truly comparative, sometimes it does use some hierarchical notions. Thus, Cuvier (1835) gives up the idea of a unique plan common to all living things that would necessarily lead to a hierarchization of living forms: "There exist not one, but four plans, four general forms according to which all animals seem to have been modeled" (vol. 1, p. 68). Nevertheless, when he is counting the digits of mammals, he distinguishes perfect digits from "imperfect rudimentary" digits (vol. 1, p. 82), which involves a hierarchical notion.

41. Natural history had not been enough to finish off the anthropocentric hierarchical vision of the world of living things. That vision was even pushed as far as possible in the 1768 work of Jean Baptiste René Robinet, *Considérations philosophiques sur la gradation naturelle des formes de l'être, ou les essais de la nature qui apprend à faire l'homme* ("Philosophical Considerations of the Natural Gradation of the Forms of Being, or the Trials of Nature Learning to Make a Man"). Thus, for Robinet, some phallus-like stones, which he calls "priapolites," were something like a sketch, a "mock-up" one might say, of the human penis. See the cutting remarks on Robinet by Cuvier (1841, vol. 3, pp. 82–83).

An accurate understanding of the extent of Aristotle's anthropocentrism should keep us from the excesses of certain anachronistic interpretations, to which I have already called attention. Despite the profound differences between Plato's zoology and Aristotle's, I cannot agree with Friedrich Solmsen (1978, 479), who claims that "while the

Timaeus is anthropocentric, Plato contenting himself with some very few and very brief glances at other living creatures, Aristotle's biology has the entire variety of animals for its subject. Man, with whom Aristotle is preoccupied in his ethical and political writings, is in biology just one of the large variety of beings. Whether nature has made special provisions for him is a question which Aristotle takes up where he meets it on his way, but it is not his primary concern." This question is treated especially well by Lloyd (1983, chap. 3).

42. Since I shall often cite names of animals and of groups of animals, I should mention that the identification of the various species named by Aristotle is often problematic for reasons that we could call accidental — especially in the cases where, because of loss of texts, we find few or no sufficiently explicit references outside Aristotle's works. But there is a more fundamental reason, the more weighty the farther we get from particular species to more general groups. Manquat, as I noted earlier, remarks that the homonymy between Aristotle's terms and ours should not fool us. Bertier correctly writes (1977, 48): "Every attempt to compare Aristotle's systematic with that of classical and modern zoology raises inextricable difficulties of terminology. We are trying to transcribe the unities of classification of the *History of Animals* into modern languages that owe much, in their outlines, to Aristotle, but not everything. Consequently, we have a hard time seeing the differences and semantic boundaries of terms that appear to be Aristotelian, but are so only by tradition." I have given several examples of mismatch in note 6 of the Introduction. Bertier advocates transliterating Aristotelian terms, or translating them literally. As my object is not the *identification* of animals and animal families, I shall usually keep the modern terms, even if they are only approximate.

43. This interpretation of the word *anonymos* is the only one possible and is confirmed by other Aristotelian usages. For example, in *HA* 1.6.490b31–491a3: "In the *genos* of animals that are simultaneously quadrupeds and viviparous, there are many *eide,* but they are anonymous: we call them, so to speak, by the name of the individual — for example, man, lion, elephant, horse, dog, and other animals of that kind — since there is no unique *genos* except in the case of animals that are called 'long-maned,' as the horse, donkey, mule, burro, and the animals that are called 'half-ass.' " As for the grammatical construction at *HA* 2.15.505b30–32, I believe that we must make *eidos* (line 31) the subject, and the predicate must include both εἴ τι ἄλλο ἀνώνυμόν, διὰ τὸ μὴ εἶναι γένος, as a causal accusative, and ἁπλοῦν. That absolutely excludes the possibility that εἶναι in line 30 has an existential sense, as Balme claims (1962a, 93).

Aristotle distinguishes two kinds of crocodiles, river and land, at *HA* 5.33.558a14. He actually describes only the river crocodile, and it is thought that he never saw one, since his description is very close to the one Herodotus gives, including the errors, particularly the mobility of the upper jaw and the flexions of the limbs. Manquat (1932, 37–45) usefully prints in parallel the texts of Herodotus and Aristotle on various animals, including crocodiles. Would the land crocodiles be large lizards, such as iguanas?

44. This solution is not, it is true, trouble free. It may be easy enough to count the crocodile as one *eidos* of the oviparous quadrupeds, but not so easy in the case of the serpents. Serpents can perhaps be counted as a limiting case of the quadrupeds (see *IA* 7.707b22), but even so they are not all oviparous, since the viper is viviparous, or rather ovoviviparous. As I noted earlier, from the perspective of the sexual organs the family of serpents must be broken up, but from the perspective of the other internal parts it can perhaps be considered an *eidos* of the *genos* "oviparous quadrupeds."

45. *PA* 4.5.680a15: οὐ γὰρ ἓν εἶδος τῶν ἐχίνων πάντων ἐστί. Louis translates: "il existe plusieurs genres d'oursins (en effet tous les oursins n'appartiennent pas à la même espèce)." Although this translation is incoherent, it does have the merit of showing us that there is a difficulty. That difficulty is purely and simply ignored by A. L. Peck in his translation: "There are several kinds of sea-urchin" (Loeb edition, p. 327). The Didot edition translation says: "Sed quum non unum, sed plura genera erinaceorum sint omnium" (vol. 3, p. 279).

46. For more details on this point, see Pellegrin (forthcoming).

47. See *PA* 4.10.686b11; *IA* 11.710b13; *PN* 453b6. Comparisons between children and animals are fairly frequent in Aristotle. Children, he says, have a "language" comparable to that of the viviparous quadrupeds (*HA* 4.9.536b5); they have, like the animals, traces of their future characteristics as adults (*HA* 8.1.588a19); they cannot, as animals cannot, serve as norms: a passage in the *Eudemian Ethics* (7.2.1236a2) says that the "pleasant," which serves as a point of reference, is that of adults and not that of children and animals (cf. 2.1.1219b5; *EN* 1.10.1100a2, 3.3.1111a21, 6.13.1144b8). Aristotle also notes the imperfection of children by comparing them to women: they have feminine forms, for the female is a sterile male (*GA* 1.20.728a17), and like women they are never bald (*GA* 5.3.784a24).

3.   THE STATUS AND FUNCTIONS OF ARISTOTLE'S ZOOLOGICAL CLASSIFICATIONS

1. Louis (1955a, 304): "Far from classifying animals in a systematic and artificial way, he often modifies his point of view in accordance

with the progress of his research. His hesitations are worthy of the true scientist that he was. Should we be surprised that he never succeeded in liberating himself completely? In any case, they represent a real interest for anyone who is concerned with the history of Aristotle's work. For it is probable that the various classifications that he adopts correspond to successive stages in the development of his thought."

2. Aristotle also notes, from Book 1 of the *History of Animals* onward, that the order of the study should sometimes be reversed because man is not the best known from all points of view: "The parts, then, that are externally visible are arranged in the way above stated, and as a rule have their special designations, and from use and wont are known familiarly to all; but this is not the case with the inner parts. For the fact is that the inner parts of man are to a very great extent unknown, and the consequence is that we must have recourse to an examination of the inner parts of other animals whose nature in any way resembles that of man" (*HA* 1.16.494b19, trans. D'Arcy Thompson). Furthermore, this passage is invoked by those who think that the dissection of human cadavers was not practiced in Greece in Aristotle's time. On this topic, Ménétrier (1930) thinks that Aristotle and the pre-Alexandrian physicians had knowledge of human anatomy from, if not dissection, at least the examination of fetuses or of cadavers of abandoned and exposed infants. From this follows, for example, the errors made by the authors of the Hippocratic corpus in regard to the morphology of the uterus and its horns, because of their smallness in young organisms, or the fact that Aristotle argues that the human kidney has lobes and is formed of several kidneys (*PA* 3.9.671b6), which is false in the adult but true in the child. Let me cite a passage that seems to show that there were dissections only of animals: "Those who say that children are nourished in the uterus by means of suckling a bit of flesh are mistaken. If this were true, the same would occur in the other animals, but it is not found to do so, as can easily be observed by means of dissections" (*GA* 2.7.746a19, trans. Peck).

3. It is well known that Aristotle had a real theory of the adaptation of living things to their environment, "for nature looks for the adapted" (ἡ γὰρ φύσις αὐτή ζητεῖ τὸ πρόσφορον) (*HA* 9.12.615a25). In HA 8.28, Aristotle notes, for the most part without explaining, the variations according to places, doubtless because, despite all his intellectual ingenuity, he did not succeed in finding a teleological explanation of the fact that in the island of Cephallenia a waterway separates the region where there are cicadas from the region where there are none. But it is true that the *History of Animals* reports facts without giving the cause (we will see this problem again): thus, at 606b17, when Aristotle asserts without justification that "in general,

fierce animals are fiercer in Asia, but all are more courageous in Europe," that does not mean that he cannot explain this peculiarity. We know that he applied it to human beings in developing a "climatic theory" (see *Politics* 7.7.1327b23), often related to the one developed in the Hippocratic treatise *Airs, Waters, Places.*

Not all variations are explained as adaptations: when Aristotle asserts, in *PA* 4.10.686b1, that "some animals have become (ἐγένετο) quadrupeds because their soul cannot support their weight," environment is not involved in the explanation.

4. Perhaps at the very end, if the brief chapter on the ostrich is interpolated as some think; but Pierre Louis is doubtless right in thinking that it is not.

5. The historical and epistemological significance of this "overlapping of groups" has been studied by G. E. R. Lloyd (1983). I cannot agree with him when he says that "in some notable instances of creatures previously considered with special interest or respect as boundary-crossers, Aristotle's conclusion is to treat them as well-established natural groups" (p. 52).

6. Joly (1968, 246): "Today's reader can only be consternated before these multiple comparisons borrowed most often from rudimentary techniques." Joly then cites a passage where Aristotle compares nature to a master of an economical household (*GA* 2.6.744b17). "From the Greek viewpoint," Joly continues, "this reference is still technical. In the *Movement of Animals,* the living body is similarly compared to a city provided with good laws: in the nineteenth century, it was by biology, at last better known, that one tried to enlighten sociology; here, in opposition, it is politics that tries to clarify biology." Even if one accepts Joly's point of view, principally insisting upon the prescientific and, as it were, infantile character of Aristotelian biology, is it so surprising that a discipline in such an embryological condition should borrow models from a more "advanced" discipline? In any case, Aristotle justifies his recourse to these advanced disciplines: thus, medicine and gymnastics allow progress in ethical speculation. "We must first consider that they [the ethical virtues] have a nature that makes them perish by lack and by excess, as we see in the case of strength and health (for we must, to understand what is hidden, use manifest witnesses)" (*EN* 2.2.1104a11). See Jaeger (1957). Preus (1979) has best clarified the range and limits of the comparison between biology and politics in relation to this passage in the *Politics.*

7. Aubonnet ed., vol. 1, p. 300, n. 11. The formulation is more or less similar in the translation by Tricot, p. 270, n. 7.

8. See Pellegrin (1981).

9. Bonitz notes only this one occurrence of this verb. To indicate

definition or the process of defining, Aristotle generally uses the terms διορίζειν, ὁρίζειν, ὁρισμός, διορισμός, ὅρος.

10. Nevertheless, a corresponding enumeration is found applied, not to the organs themselves, but to the functions, in *PA* 2.1.647a24: "The sensitive, motive, and nutritive faculties (δύναμις) are all found in the same part of the animal's body."

11. See, for example, *PA* 2.10.655b33: "For we assert that plants are alive." Is there here a trace of a polemic with people or schools that asserted the contrary? For references concerning the difference between plants and animals, see Bonitz *Index* 839b49.

12. κοιλία is well enough translated "belly": this word designates for Aristotle the stomach as well as intestines, while ἔντερον designates only the intestines. More exactly, when Aristotle wants to distinguish stomach from intestines within a given passage, he calls the first κοιλία and the second ἔντερον (e.g., at *PA* 3.14.674a11). The κοιλία is the part in which concoction occurs (see note 13, below). Sometimes Aristotle clarifies by saying ἄνω or κάτω κοιλία. At *GA* 1.18.725b1, Aristotle says that the "bottom" of the belly is reserved for the residue of solid food, the bladder for the liquid residue, and the "top" of the belly for useful nourishment.

13. To be more precise, we must say that Aristotle carries out the division of the *genos* of digestive organs in two different directions. First, there are several organs answering to the plurality and succession of nutritive functions. In the complete animals (i.e., those in which the corporeal form approaches that of man, who is the perfect animal, in opposition to less differentiated animals, e.g., some bloodless animals), there are three (*HA* 1.2.488b30; cf. *PN* 468a13): the mouth (στόμα), the belly (see note 12, above), and the part that serves for evacuation and that, Aristotle says, "has several names" (489a2) — it is difficult to say whether he is speaking of the anus alone or whether he means in addition the rectum or even a larger section of the intestines. Can we suppose that we have here a division of a *genos* into *eide*? In fact, it seems that between these organs (functions) there is a more nearly hierarchical than contrary relationship. The belly is, in fact, the essential organ for nutrition, for that is where concoction (πέψις; see *PA* 2.3.650a2 et seq.) occurs, and that is, as it were, the essence of nutrition. The function of the mouth and esophagus is subordinate to the function of concoction: in carrying the food to the belly after having divided it, if necessary, the mouth and esophagus allow an efficient concoction (they are the cause of εὐπεψία; *PA* 2.3.650a11; cf. 4.3.677b31). Excretion is also obviously subordinate to digestion proper.

Second, there is a division of each of these organs according to animal species. Here there is indeed a division of *gene* into *eide*, and that is how we must take the passage quoted above: "These parts are both same and different in respect to the ways already noted: either in respect to *eidos,* or excess, or analogy, or position" (*HA* 1.2.488b31). It is always the *function* that defines the extension of the *genos.* Thus, since the mouth has the function of preparing food for concoction in the belly, the mouth of birds is, as it were, divided into two separate parts: "Since they have nothing that really fulfills the task of a mouth (τὴν τοῦ στόματος . . . λειτουργίαν), for they have neither teeth nor anything else to cut or grind the food, some have before the belly what is called a 'crop,' which fulfills the function (ἐργασία) of the mouth, others have a wide esophagus" (*PA* 3.14.674b18).

14. All vital functions are derived from fundamental and irreducible functions that thus logically operate as true ἀρχαί. See *PA* 1.5.645b22: "By common functions (πράξεις), I mean those that belong to all animals; by those present in a *genos,* I mean those of animals whose differences we see to be differences of excess in relation to one another (for example, I speak of birds as a group, of man as a species *[eidos]*) and everything that, in general, contains no difference (μηδεμίαν ἔχει διαφοράν). For animals have common properties according to analogy, according to *genos,* and according to *eidos.* Thus, for the functions that are for the sake of other functions, it is obvious that the organs of which they are the functions are in the same relationship as those. Similarly if some functions are first and are the end of other functions."

Aristotle does not list reproduction among the fundamental functions that define the animal, although he recognizes, from a *practical* point of view, that it has an importance equal to that of the function of nourishment: "A part of the life of animals is thus devoted to actions relative to reproduction, just as another part is devoted to nutrition. In fact, the efforts, the life, of all of them is found to be concentrated on these two objectives" (*HA* 8.1.589a3). But reproduction is not definitional of the animal, since there are beings belonging to the *genos* animal that appear by spontaneous generation. Thus, when, in *HA* 8.1.588b28, Aristotle says that the actions relative to reproduction are κοιναὶ πάντων, we must not understand, as Pierre Louis did, "common to all living things," but common to those in question in the passage: plants and animals that have reproduction as a function (ἔργον, line 27). From a *theoretical* point of view, however, we may suppose that reproductive function is not for Aristotle sufficiently *autonomous* to be one of the primordial functions that define the living being. In fact, in the *De Anima,* Aristotle subordinates the generative function to the nutritive. Thus, he says of the nutritive soul: "Its functions are reproduction and the use of food; for it is the most natural function in living things, such as are perfect and not mutilated or do not have spontaneous generation, to produce another thing like themselves" (2.4.415a25, trans. Hamlyn). Cf. 416a19: "Since it is the same potentiality of the soul that is nutritive and reproductive. . . ." At 416b14 et seq., Aristotle explains that "the ensouled thing maintains its substance and exists as long as it is fed, and it can bring about the generation not of that which is fed, but of something like it." Doubtless we should see here, following in particular Themistius, a reaffirmation, in another form, of the Aristotelian understanding of seed as a residue of food.

Similarly, Cuvier (1835, vol. 1, p. 31) says that an animal is "a sensitive, mobile, digesting bag." Elsewhere he distinguishes three levels of functions: sensation and movement, nourishment, and reproduction (ibid., pp. 16–17).

15. Life, to which these various functions can be related in the complete animal, appears in the very formation of the living being, and all those functions proceed from the same principle. That is what *PA* 2.1.647a24 says: "The sensitive, motive, and nutritive faculties are found in each animal in the same part of the body, according to what has already been said in other writings. It is necessary that the part containing the origin of such principles be . . . simple and . . . anhomoiomerous. That is why in the bloodless it is the analogue of the heart; and in the blooded, the heart." Several other passages put these functions one by one or by twos into the heart: "That the principle of perception comes, in animals, from the same part from which comes their motility has been established previously in other writings; . . . in blooded animals, it is the part that surrounds the heart" (*PN* 455b34). At *PN* 474a31, the principle of the first (i.e., nutritive) soul is placed in the heart: there is perhaps an evolution here in Aristotle's position on the subject of the exact location of the vital principle.

At this point, we can note that this character of the heart as principle furnishes some evidence that may illuminate a controversial issue: must we see in Aristotle a "true scientist," as Louis says (see note 1, above), for whom observation of facts is ultimately decisive, or a resolutely prescientific thinker, whose fundamental objective is to illustrate, and to save, metaphysical principles? Wouldn't Aristotle's often repeated assertion that the heart is formed first (e.g., *PA* 3.4.665b10; *GA* 2.1.735a23, 4.738b16, 740a3; *PN* 468b28) show us, because of the embryological observations it presupposes, that Aristotle establishes these principles on the basis of observation? In that case, the metaphysical primary characteristic would be based on the chronologically first principle, as empirically observed. Simon Byl (1968) has argued that, even on the subject of the heart, metaphysics remains determinative: it is the valuing in the entire work of Aristotle, and (Byl adds) in Greek thought in general, of the "mean" *(mesotes)* that appears to lead to the affirmation that the heart is a principle, even at the price of placing it, *contrary to observation,* in the middle of the body, with (in man) a slight tendency to lean toward the left.

Though Byl's case might be sustained for Aristotle, it cannot be generalized so easily, since several of the medical and philosophical writers before Aristotle, and many after him, believed that the governing part of the soul was located in the brain. Aristotle himself shows us that this was a bitter controversy in his own day; in *Metaphysics* Zeta 10.1035b26, concerning the essential part of the animal, he says that "some parts are neither prior nor posterior to the whole, i.e., those which are dominant and in which the formula, i.e., the essential substance, is immediately present, e.g., perhaps the heart or the brain; for it does not matter in the least which of the two has this quality" (trans. Ross). We may say of this passage that Aristotle did not want to decide this highly controversial issue in the context of a metaphysical discussion.

In a brilliant article, Suzanne Mansion (1973) tries to grasp to what extent the Aristotelian concept of life could be illuminated from different angles from the "youthful" to the "mature" works, or works of "old age," by comparing the *Protrepticus* and the *De Anima.* In the earlier of these works, she writes (p. 434), "Aristotle thinks about it [life] starting from perception, and from this point of view tries to interpret all manifestations of life, perhaps without complete success." It seems that we may find a turning point in Aristotle's concept in a passage of the *Topics* (6.10.148a29), cited by Mansion in a note, but she could have given it more emphasis: ἡ δὲ ζωὴ οὐ καθ' ἓν εἶδος δοκεῖ λέγεσθαι, ἀλλ' ἑτέρα μὲν τοῖς ζῴοις ἑτέρα δὲ τοῖς φυτοῖς ὑπάρχειν ("Life does not seem to be said according to one unique *eidos,* but one finds it one way in animals, and another in plants"). The road is thus open to the three-functional schema of life of which we have spoken. Mansion shows, finally, how the contemplative life in the last works conserves the "esthetic" conception of life of the *Protrepticus.*

16. For self-nourishment of plants, see, e.g., *De An.* 2.2.413a25 et seq. For sensitivity and movement, see *PN* 436b10: "Each animal, inasmuch as it is an animal, necessarily possesses sensation, for it is thus that we distinguish what is an animal from what is not." Cf. *HA* 1.1.487b6: "Some animals remain in the same place, others move. Those that remain fixed live in the water; no terrestrial animal remains in the same place." Also see *De An.* 1.5.410b19, 3.9.432b20, and the references in Bonitz *Index* 472a32–35. Suzanne Mansion (1973, 436) speaks of "Aristotle's tacit retraction" when he finally refuses sensation to plants. If, in the *Protrepticus,* life is defined by sensation, Aris-

totle, following Plato in this respect (see p. 435, n. 23, of Mansion's article), would at first have attributed sensitivity to plants.

17. Between fixed animals and truly mobile animals, there is the intermediate *genos* of testacea. See especially *IA* 19.714b18.

18. This relation is therefore not at all one of analogy, any more than there is analogy in the Aristotelian doctrine of "being that is said in several ways" (*Metaph.* Gamma 2.1003a33). Cf. Aubenque (1978).

19. Hegel (1974, vol. 2, p. 157). Sometimes Aristotle adds the "dianoetic" function to the list of functions that define the living thing (*De An.* 2.2.413b12, 414a32; the latter passage adding also a "desiring" function). But Aristotle more or less excludes the consideration of this dianoetic function from his biological studies as such: "Concerning the intellect and the potentiality for contemplation, the situation is not so far clear, but it seems to be a different kind of soul (ψυχῆ γένος ἕτερον), and this alone can exist separately" (*De An.* 2.2.413b24, trans. Hamlyn).

20. Nor is "being" a predicate, which allows us to push our comparison between life and being a little further. A cadaver is not a living being that has lost a characteristic; a living thing and a cadaver, as Aristotle often says, are not the same except "by homonymy."

21. See Bachelard (1932, chap. 5, p. 96), who establishes, contra Wilhelm Ostwald (1908), that Mendeleev's classification is of a different type than the taxonomy of natural historians.

22. Cuvier (1835, vol. 1, pp. 48–49) thinks that the various possible combinations of organs actually exist in real animals, unless they are incompatible.

23. This is a principle of "compatibility" that is close to what we have also seen in Cuvier. See note 22, above.

24. See *Metaph.* Epsilon 1. At *PA* 2.7.653a9, Aristotle writes on the subject of "fluxes" (ῥεύματα): "But it is rather among the principles of diseases that we should deal with such things, to the extent that it is for natural philosophy (φυσικὴ φιλοσοφία) to speak to them." The question of whether or not this study belongs to natural philosophy is posed as soon as one goes on from anatomical and physiological study to medical applications. But this passage takes it for granted that biology is indeed a part of natural philosophy.

25. *PA* 1.1:

639b30   ἀλλ' ὁ
640a1   τρόπος τῆς ἀποδείξεως καὶ τῆς ἀνάγκης ἕτερος ἐπί τε τῆς
640a2   φυσικῆς καὶ τῶν θεωρητικῶν ἐπιστημῶν.

26. Peck, Loeb edition, p. 58, n. (b). This distinction between a science that nature has and a science that we have of nature seems very far from Aristotelian, especially if we remember that for Aristotle

science is a disposition of the knowing subject: ἐπιστήμη ἐστὶν ἕξις ἀποδεικτική (EN 6.3.1139b32).

27. LeBlond (1945, 137, n. 23). We should notice that despite what LeBlond says, Michael does not correct the text, and that the passages to which LeBlond refers (EN 6.9.1142a17 and Metaph. Gamma 3.1005b1) in no way exclude physics from the theoretical sciences. See Michael of Ephesus, In Libros De Partibus Animalium, Michael Hayduck, p. 3, lines 20–29.

28. We should notice that Düring's interpretation is not entirely novel, since Charles Thurot (1867), in his very valuable "Critical Observations," had already posited its principle: "It is clear that the science of nature and the theoretical sciences are opposed to the practical sciences from which Aristotle has just gotten his examples; but the science of nature is not opposed to the theoretical sciences, as has been believed" (p. 208).

29. An example of a statement that I call "continuist" would be: "Art is more nearly science than is experience" (Metaph. Alpha 1.981b8).

30. In the Posterior Analytics, 1.13, knowledge of the fact is used by knowledge of the why. But when one cannot claim to attain the scientific level, knowledge of the fact is enough: that is the case in the Nicomachean Ethics (1.2.1095b7).

31. This is the position of Louis, LeBlond, and others. The most developed form of this thesis is probably that of Düring (1943, 25–30), who claims to find in the biological corpus the trace of three successive courses given by Aristotle. The first would have included the present History of Animals, but in its primitive form as a collection of separate essays on: the parts of animals (1–4.7), generation (5–6 and perhaps 7), nourishment (8.1–11), actions of animals (8.12–17), health and disease (8.18–30), and habits of animals (9). To this first course would also belong the treatise on The Progression of Animals. The second course would have left traces in the following treatises: Resp., PA 2–4, Juv., Long. Vit., early versions of the Sens. and De An., and perhaps Mete. 4. The third course, complete in this case, would have been given by Aristotle upon his return to Athens: he would have put in order and/or revised the older treatises and added new ones to them, which would give the following sequence: PA 1, HA, PA 2–4, IA, De An., MA, PN, GA. Düring's interpretation, although certainly "chronologist," nevertheless asserts that the biological corpus as it has come down to us through the edition of Andronicus reflects Aristotle's conscious goal of systematic exposition, which is a fundamental idea of my own interpretation.

32. Balme argues this thesis in two essays not yet published: "The

Place of Biology in Aristotle's Philosophy," contributed to the collo-
quium on Aristotelian biology at Williamstown, July–August 1983, to
be published in *Philosophical Issues in Aristotle's Biology,* ed. Alan Gotthelf and James Lennox; and "Aristotle's *Historia Animalium:* Date and
Authorship," which is expected to serve, in revised form, as part of the
Introduction to the last volume of the *History of Animals* in the Loeb
collection.

33. Düring (1961). The analogical relationship of *mathematics is to
astronomy as physics is to zoology* is interesting and confirms Düring's
interpretation, in that it posits a relationship of subordination between
zoology and physics, i.e., *a fortiori* a relation of "belonging."

34. Balme, "The Place of Biology in Aristotle's Philosophy," p. 4
in the typescript (my italics). This difference between systematic and
chronological orders has been clearly indicated by Allan (1952, 97):
"Concerning these works, then, two problems arise: first, in what
systematic sequence does Aristotle intend them to be read? and secondly (in view of possible inconsistencies in doctrine), what is the
probable order of their composition?" But the author does not at all
exploit this distinction.

I will only say a few words about the *internal references* in the biological corpus, on which the partisans of the traditional chronological
ordering claim to rely. When Aristotle refers to the *History of Animals*
in the other biological treatises, it is, Louis tells us, "always in the past
tense, as to a completed work that the reader can consult" (vol. 1, p. X
of his edition of the *HA*). In this proposition there is both a truism and
an error of perspective. The truism is that when one refers to something, the object referred to generally has to exist (although there can
also be references to the future, as I shall show). The error of perspective is that, whether he wants to or not, Louis tends to consider the
History of Animals as a datable work, thus as a work written during a
short period. In the first place, there are not very many such references:
Louis counts twenty-seven to the *History of Animals* in the other treatises; and they could very easily have been added afterwards. Balme, in
fact, notes that "some of these references are reciprocal, showing that at
least one of the pair was inserted after the work was written" ("The
Place of Biology," p. 3). (Even Balme seems to think of the *History of
Animals* as a work written once.) In the second place, nothing proves
that the ἱστορία or ἱστορίαι that Aristotle cites are not a set of data, a
common source for the present form of all the biological treatises,
including the *History of Animals.* That is Balme's position, and it could
be Düring's too. Finally, there are in the other treatises references *to the
future* that seem to refer to passages in the *History of Animals.* Thus, at
PA 2.14.658b12, concerning the eyelashes, Aristotle writes: "What

can be said about them ought to be reserved for the appropriate occasion," and the "appropriate occasion" seems to be *HA* 2.1.498b21.

The most prudent solution is thus to consider the internal references as relating either to the systematic order of the treatises as Aristotle wanted them or to a *relative* chronology — that is, to recognize that this passage in this treatise is prior to that passage in that treatise, without attempting to deduce from that anything about the treatises as a whole. For example, we read in *PA* 4.10.689a17: "How the internal organs are arranged and how they differ in respect to seed and conception is clear from the *History of Animals* and the anatomical drawings (ἐκ τε τῆς ἱστορίας τῆς περὶ τὰ ζῷα φανερὸν καὶ τῶν ἀνατομῶν); later that will be discussed again in *On Generation.*" François Nuyens (1948, 158, n. 37), partisan of the traditional chronological ordering, sees here this sort of caution by Aristotle himself. The future-tense "will be discussed" (λεχθήσεται) more nearly resembles an announcement made by a professor who has already worked out, if not already written, his lectures in advance than a declaration of intention to write a not-yet-written book. Once more we find systematic order lurking in a vocabulary that seems at first sight chronological.

35. See, e.g., LeBlond (1945, 18). Also see Jaeger (1948, 329): "The *History of Animals* itself belongs in intellectual structure not to the conceptual type exemplified by the *Physics* but to the same level as the collection of constitutions. As a collection of material its relation to the books on the *Parts* and on the *Generation of Animals* . . . is exactly the same as that of the collection of constitutions to the late, empirical books of the *Politics.*"

36. Here are some of Aristotle's references to anatomical drawings: *PN* 478a28, b1; *PA* 2.3.650a31; 3.5.668b29, 14.674b16; 4.5.680a1, 8.684b4, 10.689a18; *GA* 1.11.719a10; 2.4.740a23, 7.746a15; 3.8.758a24.

37. Dirlmeier (1962, 20): "The *History of Animals,* more than any other text concerned with natural science, dispenses with the 'logoi' character that has been called into dispute. Methodological considerations are dealt with only in a very brief remark (1.6.491a7–14); there are no aporias; reports about predecessors are either listed without critical observation (e.g., 513a12, 523a17, 607a8), or observations are contrasted with observations (e.g., 563a9, 11)." This is true, and I can also agree that the *History of Animals* "has the task of providing the phenomenological presuppositions for the other treatises" (p. 22). But I propose to show something more: that the status of "phenomenological presuppositions" is not incompatible with the integration of the *History of Animals* within etiological research itself.

38. Here I only note the fact, but I think that there are *theoretical*

reasons why the *History of Animals* cannot be such a collection. I have dealt with some of them in my article, "Aristotle: A Zoology Without Species."

39. Louis, in Budé *Histoire des Animaux,* vol. 1, p. 15, n. 1.

40. Düring (1950) has studied in detail Athenaeus's citations from Aristotle's biological works. He concludes as follows: Athenaeus did not use the edition by Andronicus (or that which followed the edition by Andronicus), which was used by the Greek commentators. Athenaeus was contemporary with Alexander of Aphrodisias. One may thus trace from his citations a pre-Andronican edition of the biological corpus. Düring finally arrives at the result that "our list of quotations points to the existence of an ancient edition of the *HA* in six books, corresponding to 1–6 in our, i.e., the Andronican, edition" (p. 48). It should be noted that Athenaeus cites that work under the title "περὶ ζῴων μορίων." Paul Moraux (1951, 321) has an opinion close to that of Düring.

41. I share Balme's prudence: "It follows that we cannot draw sure conclusions about Aristotle's treatises from these quotations" ("Aristotle's *Historia Animalium:* Date and Authorship," p. 9).

42. For bile and the bile duct, see *HA* 2.15.506a20 et seq.; and *PA* 4.2 as a whole. For the use of the proposition "all animals lacking bile live long" as the premise of a syllogism, see *APo.* 2.17.99b5.

43. This methodological order of collecting differences before explaining them is that of Aristotle's *conscious* intentions and declarations. We will see in our examination of the *Parts of Animals* that in fact the teleological explanation very often precedes the observation of the differences of the parts, and thus obscures the observation itself.

44. *APo.* 2.11.94a21: τὸ τίνων ὄντων ἀνάγκη τοῦτ᾽ εἶναι.

45. Anthony Preus (1975, 45) has seen the *systematic* priority of the *History of Animals:* "The *History of Animals* is systematically first (the other works explicitly depend upon it); the bulk of the information seems to have been gathered during the period of travels, soon after the death of Plato."

Balme (1962b) believes that he can introduce a chronological ordering of the various theories of spontaneous generation. That of the *Metaphysics* would be the oldest because most stochastic; those of the *History of Animals* and *Generation of Animals,* more "law-like," would be later; and that of the *HA* would be the latest because it defends a more "materialistic" theory than that of the *GA.* Furthermore, the botanical works of Theophrastus would be placed between the *Metaphysics* and the *Generation of Animals.* For one thing, the assumption that the movement toward materialistic theories is the historical direction strongly resembles a modern *a priori;* for another thing, all such chronological interpretations are invalidated because they do not account for the etiological division of labor between the biological works, though that division of labor gives each of them its special purpose. Thus, *HA* 5 and 6 present, as Balme notes (p. 102), only "material factors"; but that is not because Aristotle had been converted to a materialism "more advanced" than his original spiritualism, but because the entire research program of the *History of Animals* is directed toward the point of view of the material cause.

46. Aristotle uses several terms or expressions like κατὰ λόγον or εὔλογος; the latter is the subject of a masterful study by LeBlond (1938).

The *Parts of Animals* often fulfills the promise of explaining correlations mentioned in the *History of Animals;* but that is not possible for all correlations. Thus, at *PA* 3.12.674a1, Aristotle is satisfied with simply noting that all polydactylous animals have an elongated spleen.

47. Aristotle does not hesitate to fall back on sophistical explanations to save his principles, as when he argues that fish that have only two fins are not an exception to the rule according to which blooded animals have four points of movement: "These swim by flexing the rest of the body for the rest of the movement, instead of two fins" (*IA* 9.709b11, trans. Preus [1981]).

48. *PA* 3.8.671a16: κολοβοῦν means "mutilate"; Aristotle uses it in the passive at 4.12.695b2. I do not agree with the version that Louis gives of this passage, according to which "this is the only point at which nature is found faulty" (p. 91); rather, μόνον should be translated "simply." Cf. A. L. Peck: "In them [turtles] the natural formation has simply been stunted" (Loeb edition, p. 269). Aristotle in fact notes a good many other "faults" in the regular (i.e., non-monstrous) productions of nature.

49. Jaeger (1948, 330): "It can scarcely be true, as has sometimes been asserted, that the *History of Animals* would be conceivable apart from the discoveries made by Alexander's expedition. The information it contains about the habits of animals at that time unknown in Greece, such as elephants, presupposes the experience of the march to India." In the first place, even if Herodotus mentions only the African elephant (3.114; 4.191), nothing proves that the Greeks did not know the Asian elephant before Alexander's conquests. In the second place, if there was a collaboration, it could not have lasted very long, given the rapid deterioration of Aristotle's relations with his former student.

50. *GC* 1.8.325a17: "Such are the reasons for which these thinkers [the Eleatics] developed these theories about truth. Certainly, according to pure reason, the universe might have been thus; but if one takes account of the facts, an opinion of that kind resembles insanity."

51. The three later books of the *Parts of Animals* use *genos* to designate the following forms of life: turtles, freshwater tortoises, men, dogs, cattle, birds, fish, parrot fish, insects, testaceans, molluscs, bees, flies, ants, ascidians, sea urchins, starfish, cicadas, razor fish (?), squid, snakes, and animals with lungs. Aristotle also talks there about γένη μέγιστα of crustaceans: spiny lobsters, lobsters, crayfish, and crabs.

52. The uses of the verb διαφέρειν to designate specific difference in the three last books of the *Parts of Animals* are: 2.5.651a20; 3.6.669a23, 12.673b14, 19, 13.674a4; 4.5.681a10, 8.683b26, 10.689a27, 690a4, 12.692b3.

Aristotle also uses διαφορά in the singular or plural at 2.2.647b17, b29, b33, 648a11, 4.651a15; 3.1.662a23, 4.666b27, 667a10, 14.674a21, 675a25, a32; 4.5.678a27, 682a10, 6.682a36, 10.689a22, b34, 12.692b3, 13.696b24, 697a15.

Specific difference is also indicated by the mention of contraries at 2.1.646a24, 647a17, 2.648a3, 8.654a10; 3.14.675a14; 4.2.677a26 and (with some interpretation) 5.679b32.

53. Pouchet (1884, 362, n. 1). I have found a similar error in Balme (1962b).

CONCLUSION

1. Foucault (1966, 277; 1973, 265). Like Aristotle, Cuvier has been the victim of a "linear" history of science. Thus, because of his fixist positions, he is often considered a "step backward" in comparison to Lamarck, his elder by twenty-five years. In fact, Lamarck, last of the taxonomists, marks the end of a world, while Cuvier, first of the "biologists," begins a new era.

2. Cuvier (1835, vol. 1, p. 16), and similarly for the following citations.

3. See Henri Daudin (1926, 48).

4. Charles Darwin (n.d., 46): "From these remarks it will be seen that I look at the term species as one arbitrarily given, for the sake of convenience, to a set of individuals closely resembling each other, and that it does not essentially differ from the term variety, which is given to less distinct and more fluctuating forms. The term variety, again, in comparison with mere individual differences, is also applied arbitrarily, for convenience' sake."

François Dagognet (1970) is critical of Cuvier—in my opinion, too severely critical. Doubtless his severity may be explained in part by his own project: "Essai méthodologique sur la taxinomie" is the subtitle of his book. But for Cuvier, taxonomy is no longer a *real* problem. For this reason, he is sometimes rather casual about his classifications. In contrast, Dagognet is right in being surprised at seeing Cuvier "obstinately falling back on a criterion, single and most archaic, 'the blood,' in order to divide animals" (p. 107). Would that be a poisoned gift from the master of the Lyceum?

BIBLIOGRAPHY

PRIMARY SOURCES

Aelian

Scholfield, Alwyn Faber, ed. and trans. (1958–59): *On the Characteristics of Animals.* 3 vols. Cambridge, Mass. (Loeb).

Alexander of Aphrodisias

Hayduck, Michael, ed. (1891): *Commentaria in Aristotelem, Metaphysica.* Commentaria in Aristotelem Graeca 1, 2. Berlin.

Aristotle

Bekker, Immanuel, ed. ([1854–71]; 1960): *Aristotelis Opera Omnia.* 5 vols. Reprint. Berlin.

Firmin-Didot, Ambroise, ed. and trans. (1848–74): *Aristotelis Opera Omnia.* 5 vols. Paris.

Loeb and Budé volumes and Oxford translations are listed separately.

Categories

Cooke, Harold Percy, ed. and trans. (1938): *Categories, de Interpretatione.* London (Loeb). (Also includes Hugh Tredennick, ed. and trans., *Aristotle: Prior Analytics.*)

Edghill, Ella Mary, trans. (1928): *Categoriae and de Interpretatione.* Oxford.

De Anima

Hamlyn, David W., trans. (1968): *Aristotle's De Anima, Books II and III.* Oxford.

Hett, Walter Stanley, ed. and trans. (1935): *On the Soul, Parva Naturalia, On Breath.* London (Loeb).

Jannone, Antonio, ed., and Edmond Barbotin, trans. (1966): *De l'âme.* Paris (Budé).

Ross, W. David, trans. (1942): *Psychology.* Oxford.

Tricot, Jules, trans. ([1934], 1959): *De l'âme.* Reprint. Paris.

De Caelo

Guthrie, William Keith Chambers, ed. and trans. (1939): *On the Heavens.* London (Loeb).

Moraux, Paul, ed. and trans. (1965): *Du Ciel,* Paris (Budé).

Stocks, John Leofric, trans. (1930): *De Caelo.* Oxford.

[De Plantis]

Forster, Edward Seymour, trans. (1913): *De Plantis.* In W. D. Ross, ed., *Aristotle: Opuscula.* Oxford.

Ethics

Rackham, Harris, ed. and trans. (1935): *The Athenian Constitution, The Eudemian Ethics, On Virtues and Vices.* London (Loeb).

Ross, W. David, trans. (1915): *Ethica Nicomachea.* Oxford.

Solomon, Joseph, trans. (1915): *Ethica Eudemia.* Oxford.

Generation and Corruption

Forster, Edward Seymour, ed. and trans. (1955): *On Sophistical Refutations; On Coming-to-be and Passing-Away.* London (Loeb).

Joachim, Harold Henry, trans. (1930): *De Generation et Corruptione.* Oxford.

Generation of Animals

Aubert, Hermann Rudolf, and Friedrich Wimmer, ed. (1860): *Aristotelis de Generatione Animalium.* Leipzig.

Louis, Pierre, ed. and trans. (1961): *De la génération des animaux.* Paris (Budé).

Peck, Arthur L., ed. and trans. (1943): *Generation of Animals.* London (Loeb).

Platt, Arthur, trans. (1912): *De Generatione Animalium.* Oxford.

History of Animals

Aubert, Hermann Rudolf, and Friedrich Wimmer, ed. (1860): *Aristoteles Thierkunde.* Leipzig.

Dittmeyer, Leonardus, ed. (1907): *Aristotelis de Animalibus Historia.* Leipzig.

Louis, Pierre, ed. and trans. (1964–69): *Histoire des Animaux.* 3 vols. Paris (Budé).

Peck, Arthur L., ed. and trans. (1965): *Historia Animalium, I–III.* London (Loeb).

———, ed. and trans. (1970): *Historia Animalium, IV–VI.* London (Loeb).

Thompson, D'Arcy Wentworth, trans. (1910): *Historia Animalium.* Oxford.

Tricot, Jules, trans. (1957): *Histoire des Animaux.* Paris.

Metaphysics

Ross, W. David, trans. (1928): *Metaphysica.* Oxford.
————, ed. and comm. (1924): *Aristotelis Metaphysica.* 2 vols. Oxford.
Tredennick, Hugh, ed. and trans. (1933): *The Metaphysics.* 2 vols. London (Loeb).

Meteorologica

Lee, Henry Desmond Pritchard, ed. and trans. (1952): *Meteorologica.* London (Loeb).

Movement of Animals, Progression of Animals

Farquharson, Arthur Spencer Loat, trans. (1912): *De Motu Animalium, de Incessu Animalium.* Oxford.
Forster, Edward Seymour, ed. and trans. (1937): *Movement of Animals, Progression of Animals.* London (Loeb, with Peck *PA*).
Louis, Pierre, ed. and trans. (1973): *Marche des animaux, Mouvement des animaux, Index des traités biologiques.* Paris (Budé).
Preus, Anthony, trans. (1981): *Aristotle and Michael of Ephesus on the Movement and Progression of Animals.* Hildesheim.

Opuscula and Fragmenta

Hett, Walter Stanley, ed. and trans. (1936): *Minor Works.* London (Loeb).

Parts of Animals

Balme, David M., trans. (1972): *Aristotle's De Partibus Animalium I and De Generatione Animalium I.* Oxford.
LeBlond, Jean-Marie, trans. (1945): *Aristote, philosophe de la vie. Le livre premier du traité sur les parties des animaux.* Paris.
Louis, Pierre, ed. and trans. (1956): *Les parties des animaux.* Paris (Budé).
Ogle, William, trans. ([1882], 1912): *Aristotle on the Parts of Animals.* Rev. ed. Oxford.
Peck, Arthur L., ed. and trans. (1937): *Parts of Animals.* London (Loeb). (Includes Edward Seymour Forster, ed. and trans., *Aristotle: Movement of Animals, Progression of Animals.*)

Parva Naturalia

Beare, John Isaac, and George Robert Thompson Ross, trans. (1930): *Parva Naturalia.* Oxford.
Mugnier, René, ed. and trans. (1965): *Petits traités d'histoire naturelle.* Paris (Budé).

Physics

Hardie, Reginald Purves, and Russell Kerr Gaye, trans. (1930): *Physica.* Oxford.

Wicksteed, Philip, and Francis Macdonald Cornford, ed. and trans. ([1929], 1957): *Aristotle: The Physics.* 2nd ed. London (Loeb).

Politics

Aubonnet, Jean, ed. and trans. (1960): *Politique, Livres I et II.* Paris (Budé).

Tricot, Jules, trans. (1962): *La Politique.* Paris.

Posterior Analytics

Mure, Geoffrey Reginald Gilchrist, trans. (1928): *Analytica Posteriora* (with Edghill, *Categoriae and de Interpretatione*). Oxford.

Rhetoric

Freese, John Henry, ed. and trans. (1926): *The "Art" of Rhetoric.* London (Loeb).

Athenaeus

Peppink, Simon Petrus, ed. (1936–39): *Deipnosophistae.* Lyon.

Diogenes Laertius

Hicks, Robert Drew, ed. and trans. (1925): *Lives of Eminent Philosophers.* 2 vols. London (Loeb).

Herodotus

Hude, Carolus, ed. (1908): *Herodoti Historiai.* Oxford.

Hippocrates

Joly, Robert, ed. and trans. (1970): *Hippocrate, Tome XI.* Paris.

Jones, William Henry Samuel, ed. and trans. (1923–31): *Hippocrates.* 4 vols. London (Loeb).

Littré, Émile, ed. and trans. (1839–61): *Oeuvres complètes d'Hippocrate.* 10 vols. Paris.

Lloyd, Geoffrey Ernest Richard, ed. (1978): *Hippocratic Writings.* New York.

Michael of Ephesus

Hayduck, Michael, ed. (1903): *In Libros De Generatione Animalium Commentaria* (attributed to John Philoponus). Commentaria in Aristotelem Graeca 14. Berlin.

———, ed. (1904): *In Libros De Partibus Animalium, De Animalium Motione, De Animalium Incessu, Commentaria.* Commentaria in Aristotelem Graeca 22. Berlin.

Wendland, Paul, ed. (1903): *In Parva Naturalia Commentaria.* Commentaria in Aristotelem Graeca 22. Berlin.

(See also under Aristotle, *Movement of Animals, Progression of Animals.*)

Plato

Burnet, John, ed. (1900–13): *Platonis Opera*. 5 vols. Oxford.

Phaedrus

Hackforth, Reginald, trans. (1952): *Plato's Phaedrus*. Cambridge, England.

Sophist

Cornford, Francis Macdonald, trans. (1934): *Plato's Theory of Knowledge*. Cambridge, England.

Timaeus

Cornford, Francis Macdonald, trans. (1959): *Plato's Timaeus*. New York.

Pliny the Elder

Rackham, Harris, ed. and trans. (1940): *Natural History*, vol. 3. London (Loeb).

Porphyry

Busse, Adolf, ed. and Latin trans. (1887): *Porphyrii Isagoge, et in Aristotelis Categorias Commentarium*. Commentaria in Aristotelem Graeca 4.1 Berlin.

Warren, Edward W., trans. (1975): *Isagoge*. Toronto.

Theon of Smyrna

Dupuis, Jean, ed. and trans. (1892): *Expositions des connaissances mathématiques utiles pour la lecture de Platon*. Paris. (Part 2 includes "Music.")

Theophrastus

Wimmer, Fridericus [Friedrich], ed. and Latin trans. ([1866], 1964): *Theophrasti Eresii Opera, Quae Supersunt, Omnia*. Photoreprint. Paris.

Hort, Sir Arthur, ed. and trans. (1916): *Enquiry into Plants*. 2 vols. London (Loeb).

Thomas Aquinas

Leonine ed. (1918): *Summa Contra Gentiles*. In *Opera Omnia*, vol. 13. Rome.

Pegis, Anton C., trans. (1955): *Thomas Aquinas: On the Truth of the Catholic Faith*. Garden City, N.Y.

SECONDARY SOURCES

Allan, Donald James (1952): *The Philosophy of Aristotle.* London.

Anton, John (1957): *Aristotle's Theory of Contrariety.* London.

Aubenque, Pierre (1962): *Le Problème de l'être chez Aristote.* Paris.

—————— (1963): *La Prudence chez Aristote.* Paris.

—————— (1978): "Les Origines de la doctrine de l'analogie de l'être: Sur l'histoire d'un contresens." In *Les Études philosophiques* 1978, pp. 3–12.

Bachelard, Gaston (1932): *Le Pluralisme cohérent de la chimie moderne.* Paris.

Balme, David M. (1962a): "Γένος and εἶδος in Aristotle's Biology." *Classical Quarterly* 12:81–98.

—————— (1962b): "Development of Biology in Aristotle and Theophrastus: Theory of Spontaneous Generation." *Phronesis* 7:91–104.

—————— (1972): *Aristotle's De Partibus Animalium I and De Generatione Animalium I.* Oxford.

—————— (1975): "Aristotle's Use of Differentiae in Zoology." Revised version in vol. 1 of *Articles on Aristotle,* ed. Jonathan Barnes, Malcolm Schofield, and Richard Sorabji. London. Earlier version in *Aristote et les problèmes de méthode.* Louvain, 1961.

—————— (forthcoming [a]): "The Place of Biology in Aristotle's Philosophy." In *Philosophical Issues in Aristotle's Biology,* ed. Alan Gotthelf and James Lennox. Pittsburgh.

—————— (forthcoming [b]): "Aristotle's *Historia Animalium:* Date and Authorship." In vol. 3 of the Loeb *HA.*

Bertier, Janine (1977): "Introduction à la lecture de *l'Histoire des animaux.*" In *Recherches sur la tradition platonicienne (Platon, Aristote, Proclus, Damascius).* Paris.

Bonitz, Hermann (1849): *Aristotelis Metaphysica, Commentarius.* Bonn.

—————— (1870): *Index Aristotelicus.* Berlin.

Bourgey, Louis (1953): *Observation et expérience chez les médecins de la Collection hippocratique.* Paris.

—————— (1955): *Observation et expérience chez Aristote.* Paris.

Brunschwig, Jacques (1963): "Aristote et les pirates tyrrhéniens." *Revue philosophique* 153(2):171–90.

Burckhardt, Carl Rudolf (1904): *Das koische Tiersystem, eine Vorstufe der zoologischen Systematik des Aristoteles.* Basel.

Byl, Simon (1968): "Note sur la place du coeur et la valorisation de la μεσότης dans la biologie d'Aristote." *Antiquité classique* 37:467–76.

—————— (1973): "Le Jugement de Darwin sur Aristote." *Antiquité classique* 42:519–21.

Canguilhem, Georges (1968): *Études d'histoire et de philosophie des sciences*. Paris.

Chaine, Joseph (1925): *Histoire de l'anatomie comparative*. Bordeaux.

Cury, G. (1960): "Comment pouvons-nous juger, aujourd'hui, la biologie d'Aristote?" In *Actes du Congrès de Lyon de l'Association Guillaume Budé (1958)*. Paris.

Cuvier, Georges (1835): *Leçons d'anatomie comparée, recueillies et publiées par M. Duméril*. 2nd ed. 8 parts in 9 vols. Paris.

——— (1841): *Histoire des sciences naturelles depuis les origines jusqu'à nos jours chez tous les peuples connus*. 5 vols. Paris.

Dagognet, François (1970): *La Catalogue de la vie: Essai méthodologique sur la taxinomie*. Paris.

Darwin, Charles (1872; n.d.): *On the Origin of Species*. London; New York (Modern Library).

Daudin, Henri (1926): *Les Méthodes de la classification naturelle et l'idée de série de Linné à Lamarck, 1740–1790*. Paris.

Desanti, Jean-Toussaint (1967): "Une Crise de développement exemplaire: La 'Découverte' des nombres irrationnels." In *Logique et connaissance scientifique*, ed. Jean Piaget. Paris.

Dirlmeier, Franz (1962): *Merkwürdige Zitate in der Eudemischen Ethik des Aristoteles*. Heidelberg.

Düring, Ingemar (1943): *Aristoteles De Partibus Animalium: Critical and Literary Commentary*. Göteborg.

——— (1950): "Notes on the History of the Transmission of Aristotle's Writings." *Göteborg Högskolas Arsskrift* 56:37–70.

——— (1961): "Aristotle's Method in Biology, note on *De Part. An.* 1.1.639b30–640a2." In *Aristote et les problèmes de méthode*. Louvain.

Ficino, Marsiglio (1576): "Letter to Francesco Cattani da Diaceto." In *Opera Omnia*, vol. 1, p. 952. Basel.

Foucault, Michel (1966): *Les Mots et les choses: Une Archéologie des sciences humaines*. Paris.

——— (1973): *The Order of Things: An Archeology of the Human Sciences*. Trans. anonymous. New York.

Fragstein, Artur von (1967): *Die Diairesis bei Aristoteles*. Amsterdam.

Gauthier, René, and Jean Yves Jolif (1958 [trans.], 1959 [comm.]): *L'Éthique à Nicomaque*. 2 vols. Louvain.

Gernet, Louis (1968): *Anthropologie de la Grèce antique*. Paris.

Gilson, Étienne (1962): *L'Être et l'essence*. 2nd ed. Paris.

——— (1971): *D'Aristote à Darwin et retour: Essai sur quelques constantes de la biophilosophie*. Paris.

Gotthelf, Alan (forthcoming [a]): "*Historia Animalium* 1.6.490b7–491a6: Aristotle's μέγιστα γένη."

———, ed. (forthcoming [b]): *Aristotle on Nature and Living Things*

(Studies Presented to David M. Balme on His Seventieth Birthday). Cambridge, England.

Granger, Gilles-Gaston (1976): *La Théorie aristotélicienne de la science.* Paris.

Grene, Marjorie (1974): *The Understanding of Nature.* Dordrecht.

Guthrie, William Keith Chambers, (1957–58): "Aristotle as an Historian of Philosophy: Some Preliminaries." *Journal of Hellenic Studies* 77:35–41.

Guyenot, Émile (1941): *Les Sciences de la vie aux XVIIe et XVIIIe siècles: L'Idée d'évolution.* Paris.

Haeckel, Ernst Heinrich Philipp August (1900): *The Riddle of the Universe at the Close of the Nineteenth Century.* Trans. Joseph McCabe. New York.

Hamelin, Octave (1920): *Le Système d'Aristote.* Ed. Léon Robin. Paris.

Hartog, François (1979): "Les Scythes imaginaires: Espace et nomadisme." *Annales: Économies, Sociétés, Civilisations.* 34:1137–54.

Hegel, Georg Wilhelm Friedrich (1974): *Lectures on the History of Philosophy.* Trans. Elizabeth Sanderson Haldane and Frances H. Simson. New York. (Originally published in 1840, Heidelberg.)

Jaeger, Werner (1948): *Aristotle: Fundamentals of the History of His Development.* Trans. Richard Robinson. 2nd ed. Oxford.

———— (1957): "Aristotle's Use of Medicine as Model of Method in His Ethics." *Journal of Hellenic Studies* 77:54–61.

Joly, Robert (1968): "La Biologie d'Aristote." *Revue philosophique* 158(6):219–53.

Kucharski, Paul (1964): "Anaxagore et les idées biologiques de son siècle." *Revue philosophique* 154(2):137–66.

Kullmann, Wolfgang (1974): *Wissenschaft und Methode: Interpretationen zur aristotelischen Theorie der Naturwissenschaft.* Berlin.

LeBlond, Jean-Marie (1938): *Eulogos et l'argument de convenance chez Aristote.* Paris.

———— (1939a): *Logique et méthode chez Aristote: Étude sur la recherche des principes dans la physique aristotélicienne.* Paris.

———— (1939b): "La Définition chez Aristote." *Gregorianum* 20:351–80.

———— (1945): *Aristote philosophe de la vie: Le Livre premier du traité sur les Parties des animaux.* Paris.

Leboucq, Georges (1946): "Alcméon de Crotone, père de la biologie." *Bulletin de l'Académie royale de Médecine de Belgique* 11:61–67.

Lee, Henry Desmond Pritchard (1948): "Place-Names and the Date of Aristotle's Biological Works." *Classical Quarterly* 42:61–67.

Lefèvre, Charles (1972): *Sur l'évolution d'Aristote en psychologie.* Louvain.

Lennox, James (1980): "Aristotle on Genera, Species, and 'the More and the Less'." *Journal of the History of Biology* 13:321–46.

Liddell, Henry George, and Robert Scott, revised by Henry Stuart Jones (1940; original ed. 1843): *A Greek-English Lexicon*. Oxford.

Lloyd, Geoffrey Ernest Richard (1961): "The Development of Aristotle's Theory of the Classification of Animals." *Phronesis* 6:59–80.

———— (1983): *Science, Folklore and Ideology*. Cambridge, England.

Louis, Pierre (1955a): "Remarques sur la classification des animaux chez Aristote." In *Autour d'Aristote: Recueil d'études de philosophie ancienne et médiévale offert à Msgr. A. Mansion*. Louvain.

———— (1955b): "Le Mot ἱστορία chez Aristote." *Revue de philologie, de littérature et d'histoire anciennes* 29:39–44.

———— (1956): "Observations sur le vocabulaire technique d'Aristote." In *Mélanges de philosophie grecque offerts à Msgr. Diès*. Paris.

———— (1967): "Les Animaux fabuleux chez Aristote." *Revue des études grecques* 80:142–246.

———— (1970): "La Domestication des animaux à l'époque d'Aristote." *Revue d'histoire des sciences* 23(3):189–201.

———— (1975): *La Découverte de la vie: Aristote*. Paris.

———— (1978): "Monstres et monstruosités dans la biologie d'Aristote." In *Le Monde grec: Pensée, littérature, histoire, documents; Hommage à Claire Préaux*. Brussels.

Manquat, Maurice (1932): *Aristote naturaliste*. Paris.

Mansion, Auguste (1945): *Introduction à la physique aristotélicienne*. 2nd ed. Louvain.

Mansion, Suzanne (1961): "Le Rôle de l'exposé et de la critique des philosophies antérieures chez Aristote." In *Aristote et les problèmes de méthode*. Louvain.

———— (1973): "Deux Définitions différentes de la vie chez Aristote?" *Revue Philosophique de Louvain* 71:425–50.

Marx, Karl, and Friedrich Engels (1939): *The German Ideology*. Trans. Roy Pascal. New York.

Ménétrier, Pierre-Eugène (1930): "Comment Aristote et les anciens médecins hippocratiques ont-ils pu prendre connaissance de l'anatomie humaine?" *Bulletin de la Société française d'histoire de la Médecine* 2930:254–62.

Meyer, Jürgen-Bona (1855): *Aristoteles Thierkunde: Ein Beitrag zur Geschichte der Zoologie, Physiologie und alten Philosophie*. Berlin.

Moraux, Paul (1951): *Les Listes anciennes des ouvrages d'Aristote*. Louvain.

Mugler, Charles (1956): "Platonica: I — *Politique* 292d, e; II — σύμμετρος chez Platon; III — La Formule mathématique πλήθει καὶ μεγέθει chez Platon." *L'Antiquité Classique* 25:20–31.

Nuyens, François (1948): *L'Évolution de la psychologie d'Aristote*. Louvain.

Ostwald, Wilhelm (1908): *Der Werdegang einer Wissenschaft: Sieben gemeinverständliche Vorträge aus der Geschichte der Chemie*. Leipzig.

Parker, Robert (1984): "Sex, Women, and Ambiguous Animals."
    Phronesis 29:184.
Pellegrin, Pierre (1981): "Division et syllogisme chez Aristote." Revue
    philosophique 141(2):169–87.
——— (forthcoming): "Aristotle: A Zoology Without Species." In
    Aristotle on Nature and Living Things (Studies Presented to David M.
    Balme on His Seventieth Birthday), ed. Alan Gotthelf. Cambridge,
    England.
Petit, Georges, and Jean Théodoridès (1962): Histoire de la zoologie, des
    origines à Linné. Paris.
Popper, Karl (1972): Objective Knowledge. London.
Pouchet, Georges (1884) and (1885): "La Biologie aristotélique."
    Revue philosophique 18:353–84, 531–57; and 19:47–63, 173–207,
    288–310.
Preus, Anthony (1975): Science and Philosophy in Aristotle's Biological
    Works. Hildesheim.
——— (1979): "Eidos as Norm in Aristotle's Biology." Nature and
    System 1:79–101. Reprinted in John P. Anton and Anthony Preus,
    eds., Essays in Ancient Greek Philosophy, vol. 2, Albany, N.Y., 1983.
Rey, Abel (1946): La Science dans l'Antiquité, vol. 4: L'Apogée de la science
    technique grecque: Les Sciences de la nature et de l'homme, les mathémati-
    ques d'Hippocrate à Platon. Paris.
Robin, Léon (1944): Aristote. Paris.
Robinet, Jean Baptiste René (1768): Considérations philosophiques sur la
    gradation naturelle des formes de l'être, ou les essais de la nature qui apprend
    à faire l'homme. Paris.
Rorty, Richard M. (1973): "Genus as Matter: A Reading of Metaphysics
    Zeta–Eta." In Exegesis and Argument, ed. Edward N. Lee, Alexander
    Mourelatos, and Richard M. Rorty. Phronesis, Supplement 1, pp.
    393–420.
Ross, W. David (1924): Aristotle's Metaphysics. 2 vols. Oxford.
——— (1949): Aristotle. 5th ed. London.
Solmsen, Friedrich (1978): "The Fishes of Lesbos and Their Alleged
    Significance for the Development of Aristotle." Hermes 106:467–
    84.
Soury, Jules (n.d.): "Anatomie et vivisection du caméléon dans Aris-
    tote." L'Intermédiaire des biologistes: Organe international de zoologie,
    botanique, physiologie et psychologie. Paris.
Stenzel, Julius (1929): "Speusippos." In Pauly-Wissowa Real-Encyclopä-
    die der classischen Altertumswissenschaft. 2nd series, vol. 3, cols. 1636–
    69. Stuttgart.
Strycker, Émile de (1932): "Le Syllogisme chez Platon." Revue néosco-
    lastique de philosophie 34:42–56, 218–39.

Thompson, D'Arcy Wentworth (1913): *On Aristotle as a Biologist.* Oxford.

Thurot, Charles (1867, 1868): "Observations critiques sur le traité d'Aristote *De Partibus animalium.*" *Revue archéologique* 1867, pp. 196–209, 233–44, 305–13; and 1868, pp. 72–88.

Wehrli, Fritz Robert (1951): "Ethik und Medizin: Zur Vorgeschichte der aristotelischen Mesonlehre." *Museum Helveticum* 8:36–62.

GENERAL INDEX

Adanson, M., 6, 76, 167n3
Adaptation, 196n3
Aelian, 168n6; *De Naturibus Animalium,* 3.13, 180n47
Alexander of Aphrodisias, 94, 189n25, 206n40
Alexander of Macedon, 151, 207n49
Allan, D. J., 204n34
Analogy, 56, 71–72, 81, 85, 88–94, 106, 127, 149, 154, 190n28, 202n18, 204n33; of animal and state, 122, 197n6
Analytics, 14, 24
Anatomical drawings, 139, 205nn 34 and 36
Anaxagoras, 55
Andronicus, 206n40
Anonymos, 194n43
Anthropocentrism, 91–101, 118, 190n30, 192n39, 193n41, 196n2
Anton, J., 186n19
Apodictic, 24, 134, 142, 148, 164, 171n4
Aquinas, Thomas. *See* Thomas Aquinas
Art and nature, 132–33
Asymbletos, 55–56, 184n9
Athenaeus, 37, 144, 206n40; 2.59d, 170n2
Atoma, 69

Aubenque, P., 71–72, 171n9, 190n28, 202n18
Aubonnet, J., 122, 124, 197n7
Axiology, 150
Axis, 133–34; of division, 28–32, 35, 103, 189n25

Bachelard, G., 47, 150, 202n21
Balme, D., 9, 12, 24–27, 53, 75–82, 88, 94, 96, 98, 102, 132, 135, 138, 141–42, 146, 171n7, 182nn 3–4, 191nn 31 and 34, 194n43, 203n32, 204n34, 206nn 41 and 45, 208n53
Beekeepers, 179n43
Bees, 46, 168n6, 178n42, 179n43, 181n47
Belly (*koilia*), 126, 198n12
Bergson, E., 112
Bertier, J., 168n6, 182n4, 194n42
Bile, 55–56, 142–43, 206n42
Bird, 27, 29, 39, 45, 71, 80–81, 88, 98, 102–3, 106
Blooded/bloodless distinction, 168n7, 170n11
Bone and cartilage, 88–89
Bonitz, H., 53, 77, 167n4, 174n16, 177n32, 190n26, 191n30, 197n9, 201n16
Bourgey, L., 46, 145, 182n3

INDEX OF PASSAGES CITED
FROM ARISTOTLE'S WORKS

Designer:	Mark Ong
Compositor:	Progressive Typographers, Inc.
Text:	10/12 Bembo
Display:	Bembo
Printer:	Braun-Brumfield, Inc.
Binder:	Braun-Brumfield, Inc.